THE BULLETIN

**Mount Desert Island
Biological Laboratory
Salsbury Cove
Maine 04672**

Volume 31  1992

TABLE OF CONTENTS

A Tribute to John Boylan . . . . i-iii
Report Titles . . . . iv-viii
Reports 1-168
Officers and Trustees 169-171
Scientific Personnel 172-176
Seminars 177-179
Publications 180-183
Author Index 184-185
Keyword Index 186-187
Species Index 187

A TRIBUTE TO JOHN BOYLAN
SCIENTIST
BULLETIN EDITOR
FRIEND

We want to commemorate here the life of our dear friend, John Boylan, who died last August at the age of 76 after a most productive, fulfilling and happy life as a physician and renal and electrolyte physiologist.

John was born in Plattsburg, New York, and was raised and schooled in Rochester where his father was an executive with the telephone company. John received his bachelor's degree from Georgetown University and his medical degree from the Long Island College of Medicine. Following an assignment to the medical corps in World War II, he settled with his growing family in Buffalo where he became Professor of Internal Medicine and Physiology. He spent several most profitable years abroad, first with R. A. McCance in Cambridge, then with Kurt Kramer in Göttingen and later in Munich associating closely with other notable German physiologists, Thurau, Deetjen, Ullrich and Gertz. John was interested in young people and used his teaching of physiology as a link between medicine and science. His scholarship was not limited to science. He was keenly interested in poetry, and he and his wife, Jean, shared a passion for Irish literature, their heroes being Joyce and Yeats. I think he liked to regard himself at least as a cousin to Blazes Boylan, the convivial rake in Ulysses. John's radiant personality has been greatly appreciated and loved by generations of medical students.

John was different from most of us in research. He was not competitive and truly loved knowledge and his own data for its own sake. His papers had hallmarks of excellence and penetration into basic physical concepts as they illuminated physiology. In 1959 Homer Smith invited him to our laboratory challenging him to prove whether gill membranes were or were not permeable to urea. Since this small molecule was almost a prototype of a highly diffusible substance and since the elasmobranch fish had a concentration of 300 mM in body fluids, this paradox posed a most interesting question in both physiology and physical chemistry. John worked in the kidney shed (before the rectal gland was dreamt of) and set up complex and awe-inspiring apparatus to perfuse fish gills at various temperatures, pressures and salinity to answer the basic question about the permeability of urea. He made major findings with respect to the permeability of not only urea but water and sodium, and with his usual modesty, published these important data only once, in the volume *Sharks, Skates and Rays*, the monograph of a symposium at Bimini published by the Johns Hopkins Press in 1967, edited by Gilbert, Mathewson and Rall. Rereading John's short and brilliantly-written chapter in this book is a rare treat for the physiologist. He concluded that the dogfish gill membrane is indeed impermeable to urea compared to other membranes such as bladder and skin (a difference of some hundred fold), and at the same time less permeable also to water and sodium. I recall very well his doing these experiments over a period of some five years. They gave him great pleasure, and his technical virtuosity was inspiring. He had a group of young physicians and technicians working with him, and it was about at this time that he began to bring fellows from Germany to work.

He did another remarkable series of experiments related to the work of Homer Smith and the writer. John showed that increasing plasma HCO_3^- in the elasmobranch by 20 fold (!) caused no change in urinary electrolyte excretion. This had important implications in our work, but rather than publish it himself, John simply gave the data to us.

John and Jean Boylan were stars of our community here in their elegant good looks, their unassuming culture, their love for their six children, and friendship which radiated to all of us. Not surprising, with his large family and large spirit, he had a big and powerful motorboat which he used to traverse Frenchman's Bay while he and Jean recited Yeats to each other. He was for many years on our Editorial Board, where his fine science and literacy was greatly appreciated.

In the early '70s John retired from active laboratory work here and then moved from the University of Buffalo to the University of Connecticut where he was professor of medicine. In 1979 he became Chief of Staff at the Newington Veterans Hospital in Connecticut. In 1979 he embarked on a new career which wed his close friendship with German scientists and his interest in medical education. He organized an exchange program for medical students between German and American schools which has involved by now several hundred young people with exchanges in both directions. His influence, therefore, will continue to live in the hearts and minds of these people, many of whom he took into his home and laboratory and by example gave his gentle and thoughtful character as model for their lives.

Alas, we do not only mourn for our friend but a certain way of life that is passing by. We refer to that of the true physician/scholar and the humanist scientist, who bridges with pleasure the usual gaps between science and medicine on the one hand and also science and humanities on the other.

Instead of the usual requiem, "We shall not see his like again." I would rather end by saying that from his example we hope indeed that we shall see his like again.

Thomas H. Maren

     25   25   25   25   25   25   25   25

Twenty-five years ago I made my first call in the United States via a public phone, trying to explain to my new adviser and tutor, John Boylan, Professor of Medicine and Physiology in Buffalo, that I had arrived in Washington and in a little while would be in Buffalo. This was our first direct contact. So far, he only knew me by correspondence. Thus he probably suddenly realized that my English was not as fluent as that in my letters, which had been helpfully polished by an Australian friend. Due to his patience and special ability to cope with unexpected situations, my stubborn English, and despite the nervousness of the operator demanding more dimes in the pay phone, our initial problem was overcome. It was at that time that we started a dialog which has continued ever since. As John Boylan would say, "Hilmar, there are certain things in life which grow on you." I began to realize that this man had a secret "...ein Geheimnis..." which was worthwhile looking at. After studying John Boylan for so many years, I think I can summarize this secret—his goal was to increase his knowledge in order to enrich his wisdom, as opposed to the modern trend where all too often we are rich in knowledge but poor in wisdom.

I discovered another of John Boylan's secrets—in order to achieve the enrichment of wisdom through a scholarly approach, he showed almost no competitive attitude. He didn't need to because he had already his own identity. Thus he could relax and not be offended by other opinions or life styles. Because of this attitude, a Catholic American with a strong Irish background felt comfortable with a non-Christian German of Prussian background. We collaborated not only in research but also in a struggle to find the right paragraphs and translation in Homer's Iliad and Odyssey.

After some time, need for an exchange between the U.S. and the former West Germany became clear to both of us, and we began our venture in 1979. Since that time over 600 participants have been able to learn more about themselves and about a different culture. The goal remains that of fostering understanding between North America and Europe while also fostering academic careers in the biomedical sciences. It is a two-way bridge, with the exchange to date about 2:1 of German to American students or young scientists. At John's retirement Dr. Robert Massey, the former Dean of the University of Connecticut, School of Medicine, in Farmington, assumed his role. The site of the U.S. office of the Exchange Program is the Mount Desert Island Biological Laboratory in Salsbury Cove.

A focus on what John Boylan described as the RED THREAD "den roten Faden" of each of us. "What does that mean, the red thread? One recognizes in an idea or a situation some affinity with one's self. The origin of this affinity is often lost in your history; it may even be genetic. A note is struck and something in you responds. You recognize the thread and you follow it." And "No matter what plans you had or what you anticipated or what you were determined to do, the greatest benefit to you will be unexpected. What will be worth remembering in your life, as a result of this experience is at the present time completely unknown." There is a thread backward and forward. The thread backward for John Boylan was that he finished training in internal medicine, had finished with the army, had finished a year of pathology and had taken a position as director of an outpatient department in a university hospital. Here he wanted to teach medicine to medical students. He started by teaching physiology because, "The perceived need was to supply a bridge for the student between pre-clinical and clinical years," and "In the rediscovery of the fascination of physiology and how it relates to medicine, I found my place."

In 1953 he met Dr. McCance from Cambridge and went some years later to England with a special fellowship award from the NIH and "... with five small children and a Ford station wagon." Another trip to Europe, again with his family, was to the Institute of Physiology in Göttingen. The chairman, Dr. Kurt Kramer, Professor of Physiology, became one of his closest friends. Later on, he visited Kurt Kramer in Munich on several occasions as visiting scientist.

The "red thread has several branches." We are hearing more of these branches from others. One which I would like to emphasize, however, is the warm hospitality which I always experienced. This is probably the strongest branch, going directly to Jean Boylan, his wife and mother of their six children.

Hilmar Stolte

****

A C K N O W L E D G M E N T S

The Mount Desert Island Biological Laboratory is indebted to the National Institutes of Health for substantial support. Contributions to operating costs have greatly improved the efficiency of research activities. The individual research projects at the Laboratory are funded by various private and government agencies; all of these projects have benefited from NSF and NIH grants to the Laboratory.

The Laboratory also is indebted to the American Heart Association's Maine Affiliate for its research support and to the Lucille P. Markey Charitable Trust, The Pew Charitable Trusts, The Burroughs Wellcome Fund, The Hearst and Grass Foundations and the Maine Community Foundation for Blum-Halsey Scholars for their support of research fellowships for young investigators and students.

REPORT TITLES

REINVESTIGATION OF REGIONAL DIFFERENCES IN BURSTING STRENGTH OF CLEAVING SAND DOLLAR (ECHINARACHNIUS PARMA) EGGS

R. Rappaport and Barbara N. Rappaport
Mount Desert Island Biological Laboratory
Salsbury Cove, ME 04672

Cytokinesis in animal cells is caused by physical activity of the cortical cytoskeletal region located immediately beneath the plasma membrane. A better understanding of possible regional differences in the physical properties of the cortex could reveal some aspects of the formation of the division mechanism. In 1922, E. E. Just (Am. J. Physiol.61: 505-15) concluded that the polar surfaces of spherical sand dollar eggs were mechanically weaker than the future furrow region immediately before cleavage, because eggs immersed in dilute sea water at that time "invariably" ruptured over the poles. These observations have been accorded considerable theoretical significance. Just did not remove the extracellular hyaline layer from the eggs and, because it could have contributed to the apparent strength of the surface, we repeated his work and compared the cell's behavior in the layer's presence and absence.

A conical chamber containing 40% sea water, 60% fresh water was mounted on the stage of an inverted microscope. Ten to twelve min before anticipated time of cleavage, a suspension of eggs in a few μl of normal sea water was pipetted to the surface of the water in the observation chamber. The eggs fell through 17 mm of dilute sea water to the chamber bottom where their subsequent behavior was videotaped. Only those eggs in which the orientation of the mitotic apparatus (MA) and the point of rupture could be clearly seen were included in the data. The point of rupture was marked on a previously drawn cell profile that was divided into polar, subpolar, subfurrow and furrow regions. Fertilized eggs with intact jelly, fertilization membranes and hyaline layer, and fertilized eggs from which the jelly, membrane and hyaline layer were removed by treatment with 1.0 M glycine were studied.

Contrary to Just's findings, we observed that ruptures were not localized in any particular region. When the eggs with or without jelly and membranes were immersed in dilute sea water, the percentage of ruptures in the surface regions roughly approximated the percentage of the total surface that the regions comprised. The furrow, subfurrow, subpolar and polar surfaces comprise, respectively, 18, 33, 26 and 23% of the total egg surface. In 131 eggs with intact extracellular layers, the regional distribution of rupture points was, in percentages, furrow: 27%, subfurrow: 40%, subpolar: 18%, and polar: 16%. In 79 eggs lacking extracellular material following glycine treatment and immersion in 50% calcium free sea water, the distribution of rupture points listed in the same order was 22%, 26%, 22% and 29%.

These data indicate no localized region of cortical weakness immediately before division. We speculate that Just's results may have resulted from his method and timing of observations (which were not clearly described). We found that the outflow of cytoplasm that followed rupture caused internal cytoplasmic rearrangement that resulted in reorientation of the MA so that its axis was parallel to the direction of outflow. Shortly after rupture, the pole of the MA was repositioned close to the point of outflow. To someone who did not observe the instant of rupture, the outflow would appear to have originated at the pole of one of the asters.

This work was supported by National Science Foundation Grant DCB-8903341.

THE EFFECT OF HEXYLENE GLYCOL ON CONTRACTILE RING BEHAVIOR IN DETERMINATE EGG DEVELOPMENT OF ILYANASSA OBSOLETA

Abigail H. Conrad[1], Andy Stephens[2], and Gary W. Conrad[1]
[1]Division of Biology, Kansas State University,
Manhattan, KS 66506
[2]Southwestern College, Winfield, KS 67156

Our purpose is to understand the role of microtubules before and during cytokinesis, that phase of cell division in which the cell cytoplasm is divided, equally or unequally, between two daughter cells. When animal cells divide, they form a "contractile ring" (CR), comprised largely of F-actin microfilaments, in a band of cortical cytoplasm between the two newly formed daughter cell nuclei, in a plane perpendicular to the microtubular spindle. Formation of the spindle precedes formation of the CR; during cytokinesis, the cleavage furrow rapidly constricts around the spindle, compressing its component microtubules into a "midbody," a structure that may play a role in completing the final severing of the cleavage furrow neck.

At the time of first cleavage in the fertilized eggs of Ilyanassa obsoleta (Nassarius obsoletus), a common marine mudsnail, a second CR forms in the cell cortex at right angles to the cleavage furrow. This second ring, the polar lobe constriction, constricts the cell slowly during nuclear division (Phase I of constriction), and then at a sharply increased rate (Phase II) during the constriction of the actual cleavage furrow. Both CRs constrict until only very narrow necks of cytoplasm remain; the CR of the cleavage furrow then cuts through its neck completely, whereas the CR of the polar lobe constriction relaxes (thereby allowing a major portion of the cytoplasm of the fertilized egg to become part of only one of the two blastomeres). At the time when the polar lobe CR is maximally constricted, it encircles very few polymerized microtubules in the neck. In contrast, at the same time, in the same cell, and only a few micrometers away, the CR of the cleavage furrow encircles many microtubules (Conrad et al., 1991. J. Exp. Zool. (in press)).

We have demonstrated previously that treatment of Ilyanassa eggs with taxol causes microtubules to be present in high numbers in the polar lobe neck (as well as in the spindle) and to cause the polar lobe constriction to remain constricted, often resulting in the complete cleavage of the neck. To determine if these results were effects peculiar to taxol or, instead, arose from the experimentally-induced increase in the number of microtubules in the lobe neck, we treated Ilyanassa eggs with hexylene glycol, a compound that stabilizes microtubules, but by a mechanism distinct from taxol. Results indicated that, as with taxol, polar lobe necks that form in the presence of hexylene glycol fail to relax and, instead, proceed to cleave through the lobe neck, as would a cleavage furrow. Control and hexylene glycol-treated eggs were fixed for transmission electron microscopy and are being compared at present for their relative microtubule numbers in the polar lobe neck (as well as the spindle). These results suggest that two distinct, microtubule-stabilizing drugs both cause a constriction that would normally relax, instead, essentially to become a cleavage furrow, i.e., a positive, stimulatory result that may involve a causal relationship with the presence of microtubule bundles within the constriction. (Research supported by grants from NASA BioServe [NAGW-1197] and NSCORT (NAGW 2328].

SPREADING OF SEA URCHIN (<u>STRONGYLOCENTROTUS DROEBACHIENSIS</u>) COELOMOCYTES: DYNAMICS OF THE ACTIN CYTOSKELETON AND THE EFFECTS OF ELEVATED INTRACELLULAR CALCIUM.

John H. Henson and David Nesbitt
Department of Biology, Dickinson College, Carlisle, PA 17013

Cell shape changes in the majority of cell types are mediated by the actin cytoskeleton (actin filaments and the associated actin binding proteins). The elaborate actin cytoskeleton of the sea urchin coelomocyte provides an excellent model experimental system for studying the molecular mechanisms underlying these shape changes. Coelomocytes play important roles in coelomic fluid clotting and the phagocytosis of foreign matter, and can be induced to synchronously undergo an actin-mediated shape change from a lamellipodial to a filopodial form. Although several studies have focused on filopodial formation in coelomocytes (Edds, 1977. J. Cell Biol. 73:479-491; Otto and Bryan, 1979. Cell 17: 285-293; Henson and Schatten, 1983. Cell Motil. 3: 525-534; Hyatt et al., 1983. Cell Motil. 4: 57-71), very little is known concerning the generation of lamellipodia from preexisting filopodia. The generation of lamellipodia has a general significance in cell motility since they are the primary organelles of locomotion in most tissue cells. Therefore, we have focused on the examination of the dynamics of actin and myosin II organization during the filopodial to lamellipodial shape change exhibited by spreading filopodial cells. This shape transformation involves the generation of lamellipodia and the dismantling of filopodia. The importance of intracellular calcium levels in this process was also addressed in experiments involving the treatment of spreading cells with the calcium ionophore A23187. The studies involved a combination of high resolution, digitally enhanced video microscopy, the fluorescent detection of filamentous actin and the immunofluorescent detection of myosin II.

For observation of the substrate induced filopodial to lamellipodial shape change, lamellipodial coelomocytes were collected from coelomic fluid in a low calcium anticoagulant, isolated by means of centrifugation onto a sucrose cushion and maintained in an isotonic coelomocyte culture medium (CCM = 0.5 M NaCl, 2.5 mM MgCl$_2$, 1 mM EGTA and 20 mM HEPES pH 7.4). The cells were then transformed to the filopodial form by reducing the NaCl concentration of the CCM to 0.3 M. The filopodial cells were allowed to settle onto a poly-L-lysine coated coverslip and the process of lamellipodial generation and filopod disruption was observed. For video microscopic observation the coverslips were mounted onto a slide by means of a vaseline well which had spaces for the perfusion of solutions. The cells were viewed with a Nikon 60 X (N.A. 1.4) planapochromatic phase contrast objective lens and video images collected using a MTI newvicon camera. The video signal was digitized by a Hamamatsu Argus 10 digital image processor which allowed for digital contrast enhancement, frame averaging and background subtraction. Video enhanced images were recorded on a Mitsubishi super VHS editing VCR. For the staining of filamentous actin, cells were treated with a fixation solution consisting of 0.5% glutaraldehyde plus 100 mM lysolecithin in CCM and then stained with rhodamine phalloidin (RdPh). For the localization of myosin II, cells were fixed with 3% formaldehyde, 0.1% glutaraldehyde in CCM, postfixed in 100% acetone at -20°C, and then stained with a polyclonal antiserum raised against the heavy chain of sea urchin egg myosin II followed by a fluorescein conjugated secondary antibody.

Video enhanced phase contrast microscopy of the spreading process (see figure 1) indicated that lamellipodia arise near the cell center and spread radially out to engulf the substrate attached filopods. RdPh staining of spreading cells (fig. 1 E) indicates that the lamellipodia are filled with a network of actin filaments. With respect to filopodial dissolution, many filopods appear to unravel along points distil to the advancing edge of the lamellipod (arrow in fig. 1 A), while others appear to remain as intact phase dense tracts within the cytoplasm. RdPh staining suggests that these phase dense tracts consist of persistent filopod derived actin bundles. This was confirmed in experiments in which the same cell was video taped and then fixed and stained with RdPh. This type of procedure allowed for the direct correlation between live cell behavior and the organization of filamentous actin.

The process of lamellipodial spreading and the dismantling of filopodia can be completely reversed by treating cells with 10-20 μM of the calcium ionophore A23187 (fig. 1 C, D). The ionophore-mediated elevations in intracellular calcium cause the cessation of lamellipodial spreading, the formation of actin bundles at the periphery and within the cytoplasm, and the eventual elongation of the peripheral filopodia. Internal bundles of actin filaments are evident as phase dense tracts within the lamellipodial cytoplasm (fig. 1 D). RdPh staining allowed for a comparison between the filamentous actin distribution in spread cells in the presence or absence of ionophore treatment (fig. 1 E, F). This staining indicated the dramatic rearrangement of actin which occurs in cells exposed to elevations of intracellular calcium. The peripheral network of relatively short filaments present in the lamellipodia of spread cells is transformed into bundled arrays of long filament. The ionophore mediated actin reorganization takes place only in calcium supplemented (1 mM) CCM, suggesting that intracellular calcium stores alone are not mediating this effect.

The dismantling of the filopod core bundles of actin filaments, the generation of lamellipodia and the bundling of actin in response to ionophore treatments all would be expected to involve the action of actin binding proteins (ABP) in the coelomocytes. Filament severing proteins would be expected to play a part in dismantling of filopodia, while filament cross linking proteins might be present in the actin filament networks within lamellipodia. The ABP complement of these cells is relatively unknown, particularly in comparison to the wealth of information available on the ABPs present in the sea urchin egg and early embryo. Fascin is an ABP which is known to bundle actin within the microvilli of sea urchin embryos and the filopodia of the coelomocytes. However, this bundling process shows no calcium sensitivity in vitro suggesting that the ionophore elicited effects in coelomocytes are not being mediated entirely by this protein. One major ABP which exhibits calcium sensitive associations with actin in many cells is myosin II. We have raised a monospecific polyclonal antiserum against sea urchin egg myosin II heavy chain (see western blot in fig. 2) in order to begin immunofluorescent localization studies using this probe. In preliminary double label experiments utilizing RdPh and anti-myosin, (fig. 2 A, B) myosin is seen to codistribute with actin particularly at the cell periphery and along stress fibers. We are currently in the process of labeling spreading cells with this antiserum in an attempt to determine if myosin is crucial for lamellipodial advancement or if it is involved in the selective stabilization of filopodia.

Figure 1:   Video enhanced phase contrast microscopy of a spreading filopodial coelomocyte treated with calcium ionophore (panels A - D), and rhodamine phalloidin (RdPh) staining of filamentous actin in spread cells (panels E and F). Panel A: This cell has already spread considerably. Note that lamellipodia may spread from the cell center or arise directly from a filopod (arrow). Panel B: The cell center derived lamellipodia have advanced and fused with those derived from the dismantling of filopods. Phase dense tracks in the cytoplasm mark the presence of former filopods.The arrow indicates a lamellipod which is actively spreading. Panel C: Immediately following treatment with the calcium ionophore the leading edge of the formerly advancing lamellipod (arrow) acquires filopodial extensions. Panel D: After extended exposure to the ionophore the cytoplasm of the cell becomes crisscrossed with phase dense tracks and the remaining filopodia have undergone significant elongation. Panel E: RdPh staining of spread cell showing actin network in the lamellipodia and the filopod derived actin bundles in the cytoplasm. Panel F: RdPh staining of ionophore treated cell showing the extensive cytoplasmic actin bundles and the absence of the actin filament networks.   Magnification = 1,000 X. Time in minutes and seconds is given in the upper right of each panel.

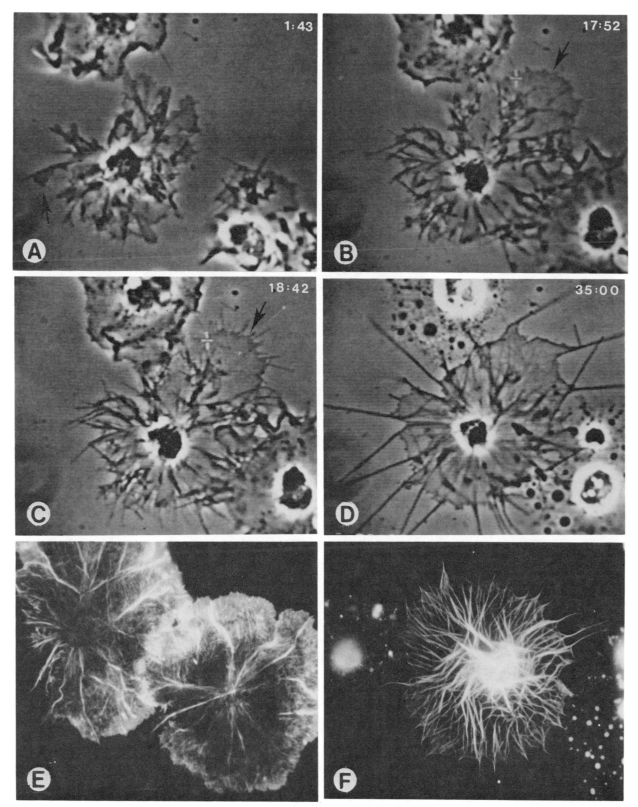

FIGURE 1

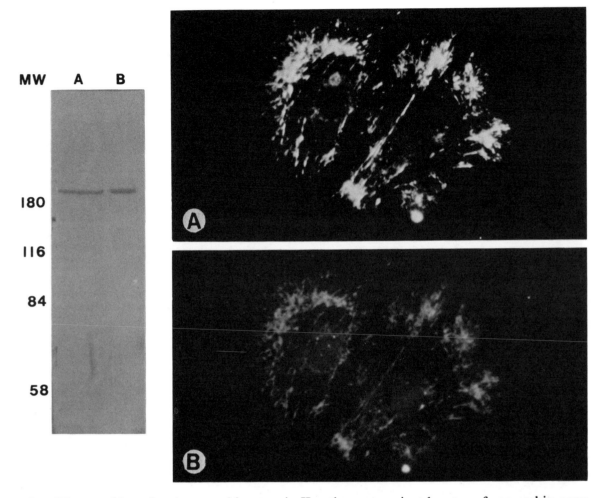

Figure 2: Western blot of anti-sea urchin myosin II antiserum against lysates of sea urchin eggs (lane A) and sea urchin coelomocytes (lane B). The single immunoreactive species present in the blots migrates at the appropriate molecular weight (200 kDa). Panels A and B: Double label of coelomocytes with RdPh (A) and anti-myosin II (B) reveals that the two proteins appear to codistribute particularly at the cell periphery, in filopods and in stress fibers. Magnification = 1,000 X.

The preliminary results reported here indicate that the process of spreading of filopodial coelomocytes involves a dramatic rearrangement of this cell's extensive actin cytoskeleton. This process offers a wealth of opportunities for examination of the reorganization of actin filaments and the in vivo functions of actin binding proteins. Future experimentation will involve the immunolocalization of major actin binding proteins during this process as well as the correlation of cell spreading behavior with ultrastructure as revealed by whole mount negative stain transmission electron microscopy. The results of these studies should provide insights into how actin mediates motility and shape changes in a variety of cell types.

Supported by a Fellowship from the Lucille P. Markey Charitable Trust and by grants from the Whitaker Foundation of the Research Corporation (C-2827) and the National Science Foundation ILI program (USE - 9050842).

HISTOLOGICAL SIMILARITIES BETWEEN GILL CHLORIDE CELLS
AND HUMAN ONCOCYTOMA CELLS

Marvin Murray [1], Jose Zadunaisky [2], Dawn Roberts [3]

[1]Department of Pathology, University of Louisville School of Medicine
Louisville, KY 40292
[2]Department of Physiology, New York University Medical Center,
New York City, NY 10016
[3]Oglethorpe University, Atlanta, Georgia 30319

Oncocytomas are unusual and uncommon adenomas of the kidney and other secretory organs. A significant observation has been the association of hypertension with renal oncocytomas and relief following excision (Stefenelli, et. al., Clinical Nephrology, 23, p.307, 1985). Oncocytoma cells are unique in that they are large, mitochondria-rich cells which appear to have the capacity for high metabolic output, but have no ostensible function. Electron microscopy of tumor cells reveals invaginated cytoplasmic membranes, microvilli, ill defined desmosomes and large numbers of mitochondria (Johnson, J.R., et. al., Urology, Vol. 14, #2, p.181-185). Chloride cells of the gills and of the opercular epithelium of fish appear to be identical to oncocytoma tumor cells in structure at the light and electron microscopic levels. The chloride cells have a specific and important function in the secretion of chloride when the fish moves from fresh water to sea water (Zadunaisky, Fish Physiol. Vol. 10B, 1984).

The similarities of structure between the two cells and the occurrence of oncocytomas in secretory organs, such as the salivary gland, the pancreas, the thyroid, lachrymal glands and kidney suggest a possible similarity in function of these two groups of mitochondria-rich cells. In this paper, histological evidence is presented to substantiate this anatomical similarity. Chloride cells from the euryhaline fish Fundulus heteroclitus adapted to fresh water, sea water, and 2x sea water were compared to oncocytoma cells of a renal oncocytoma tumor. Sections of the gills, opercular membrane, heart and kidney of Fundulus heteroclitus were also studied by conventional methods, light microscopy.

Fundulus heteroclitus caught in estuaries were transferred to sea water and used after 4-6 weeks - other specimens were acclimated to twice sea water by gradually increasing the salt content. Three groups of ten fish each were used for the experiment. Group 1 remained in fresh water with minimal salt concentration, Group 2 was acclimated in sea water, and Group 3 was acclimated in twice sea water.

Fish were pithed and specimens of gill arch, opercular membrane, heart and kidney were dissected out.

Specimens of oncocytoma were kindly provided by Dr. William M. Carney, Elizabethtown, Kentucky. The patient was a 75 year old male who developed hypertension and a mass in the kidney. The kidney was removed and the hypertension was subsequently relieved. Specimens of oncocytomas were provided preserved in 10% formalin and histologic specimens were prepared utilizing hematoxylin and eosin stain.

Exposure of the fish to increasing concentration of salt produces the proliferation of chloride cells in the gills and in the opercular membrane. Figure 1. (Karnaky, K.J., Jr., J. Exp. Zool., 199, p.355, 1977).

Sections of the oncocytoma showed typical sheets of mitochondria-rich cells with no remaining architecture nor evidence of dedifferentiation. Figure 2.

The oncocytoma cells and the chloride secreting cells from both the gill arch and the opercular membrane were similar cytologically. Similarily the several mitochondria-rich cells of the killifish (pseudobranch mitochondria-rich cells and kidney proximal tubule cells) share cytologic characteristics in common with oncocytoma cells. The killifish cell types are involved in salt and water and perhaps acid-base balance. The oncocytoma cell may share some of these functions.

The concurrence of hypertension with renal oncocytomas raises important questions. There is no report of these tumors being associated with excess renin production. Do oncocytoma cells have the potential to release vasoactive moieties?

Finally, there is the additional possibility that oncocytoma arise from mitochondria-rich cells which are functional and have preserved some of the archaic activities of those in the lower vertebrates.

Dr. Jose Zadunaisky's work was funded in part by NIH grant 01340. University of Louisville Graduate School and Department of Pathology helped fund the work of Dr. Marvin Murray. Dawn Roberts was a fellow of the Pew Foundation and the Grass Foundation.

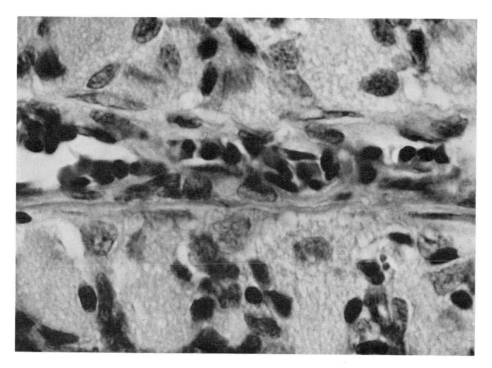

Figure 1. Proliferation of chloride cells in the gill of <u>Fundulus heteroclitus</u>
acclimated to 2x sea water. 1000x.

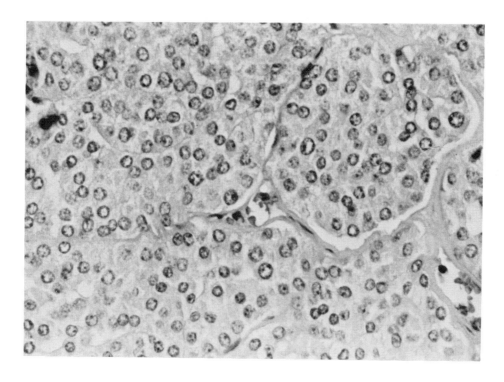

Figure 2. Renal cell oncocytoma (human). 320x.

PROLIFERATION OF CHLORIDE CELLS IN 2X SEA WATER
A PROBABLE CAUSE OF DEATH IN FUNDULUS HETEROCLITUS

Marvin Murray[1], Jose Zadunaisky[2], Dawn Roberts[3]

[1]Department of Pathology, University of Louisville School of Medicine, Louisville, KY 40292
[2]Department of Physiology, New York University Medical Center, New York City, New York 10016
[3]Oglethorpe University, Atlanta, Georgia 30319

Renal function in the fish is divided between the kidney (pronephros, mesonephros) and the gills. In a fresh water environment, the kidney has the role of conservation of electrolytes, maintenance of osmolarity and acid-base balance. In a salt environment, the fish has the need for the secretion of electrolytes in order to maintain the osmolar balance and the need to maintain acid base balance. This ordinarily occurs at the level of the gills through the function of chloride cells (Zadunaisky, Fish Physiology, p.129, 1984).

In this experiment Fundulus heteroclitus caught in estuaries were transferred to sea water and used after 4-6 weeks. Other specimens were acclimated to fresh water by gradually reducing the salt content of the water within 10 days. Three groups of 10 fish each were used for the experiment. Group 1 was kept in fresh water, Group 2 was acclimated to sea water and Group 3 was acclimated to twice sea water.

Individual fish were pithed and dissected. Each fish was weighed and each heart was weighed.

Weights of the fish and their hearts were recorded and averaged. The ratio of heart weight to total weight of the fish remained constant in the three groups studied, fresh water, sea water, and 2x sea water. This indicated that within the 6 week period of the acclimatization, there was no indication of extra work being done by the heart muscle. The histologic examination of the hearts demonstrated no evidence of hypertrophy in the individual fibers.

Histological examinations of the gill arches demonstrated the usual comb-like structure of the gill filament with fine lamellae extending at right angles (Figure 1). The lamellae contained capillaries and a procession of red blood cells involved in gas exchange. Mitochondria rich cells were seen at the base of the filament described. In sea water, the gill structure of the fish remained fundamentally the same, however chloride cells proliferated between the fine filaments and obstructed to some degree the access of the capillaries to the surrounding sea water.

In 2x sea water, the gill structure was similar to those of sea water but with a greater degree of proliferation of chloride cells. Chloride cells proliferated to the degree that the gills became clubbed, (Figure 2), thus preventing access of the capillaries to the sea water for gas exchange. Therefore, the fish was suffering from chronic obstruction of blood flow to the gill. It had been noted that many of the fish being acclimated to twice sea water died. The pathologic conclusion was that the fish died of an analog of chronic obstructive pulmonary disease, that is to say, a blockade of the capillaries preventing gas exchange, thus asphyxiating the fish. All of the gills were studied and there was obstruction of a majority of the lamellae leaving at most a fraction to be exposed to the ambient 2x sea water (Figure 3).

In the euryhaline fish, Fundulus heteroclitus, the protection of osmolar balance (plasma chloride level) is maintained even at the expense of respiration. Where high concentrations of salt exist in solution, the fish produces chloride cells until gill capillaries are obstructed with consequent asphyxiation and death.

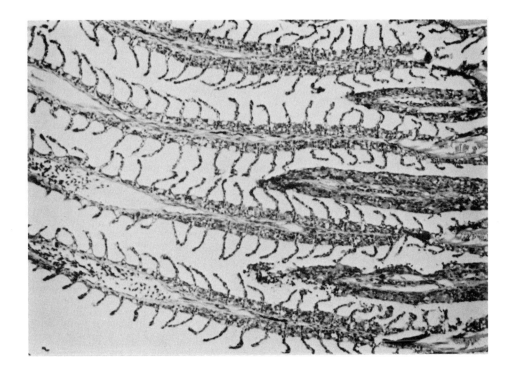

Figure 1. Gills of the _Fundulus heteroclitus_ acclimated to fresh water demonstrating freely available lamellae. 180x.

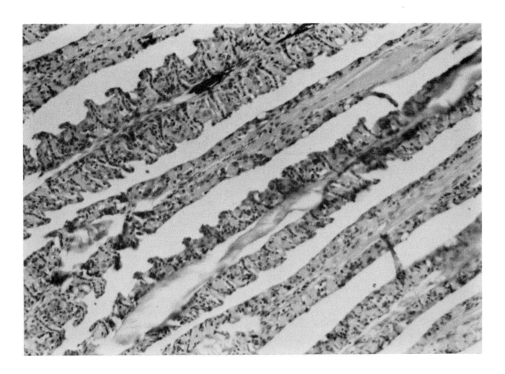

Figure 2. Gills of the _Fundulus heteroclitus_ acclimated to 2x sea water. Lamellae are sequestered from the ambient sea water by a profuse proliferation of chloride cells. 240x.

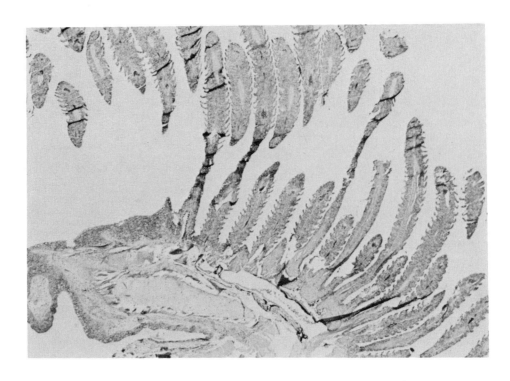

Figure 3. An entire gill segment demonstrating proliferation of salt cells and obstruction of lamellae. 30x.

Dr. Jose Zadunaisky's work was funded in part by NIH grant 01340. University of Louisville Graduate School and Department of Pathology helped fund the work by Dr. Marvin Murray. Dawn Roberts was a fellow of the Pew Foundation and the Grass Foundation.

ANOXIA AFFECTS THE MEMBRANE CYTOSKELETON OF FLOUNDER (PSEUDOPLEURONECTES AMERICANUS) ERYTHROCYTES

George Booz[1], Robert Shetlar[2] and Alison Morrison-Shetlar[2]
[1]Geisinger Clinic, Weis Institute for Research Danville, PA 17822 and
[2]Max-Planck-Institut fuer Systemphysiologie, W-4600 Dortmund 1, Germany

As for many cells, e.g. muscle, glucose (Glc) transport by flounder red blood cells (RBC) is a slow process, which is enhanced by anoxia, vanadate, and Ca^{2+} ionophore [Booz and Kleinzeller, Bull. MDIBL 29:52, 1990]. The basis for such non-hormonal increases in Glc transport is unknown. In this study, we asked if anoxia also affected the protein composition of flounder RBC plasma membranes (PM), as such an action might offer a clue as to how anoxia enhances Glc transport.

Blood from the caudal vein was centrifuged to remove white cells and platelets, and RBC resuspended to a 5-10 % hematocrit in flounder Ringer's (MOPS buffer pH 7.8) with 1 mM Na-acetate. RBC were made anoxic by incubation (12°C) under N_2 for 2 h with 1 mM KCN. Control cells were incubated under air. RBC were then spun 5 min at 2000 x g, and resuspended to 30-50 % hematocrit. PM were prepared by adapting the Watts and Wheeler method [Biochem. J. 173:899, 1978]. All steps were done on ice or at 4°C. All solutions contained 0.1 mM phenyl-methyl-sulfonyl-flouride (PMSF). While vortexing, a 10-fold volume of buffer A (in mM: NaCl, 3; $MgCl_2$, 2; Tris, 8.5; pH 7.8) was added to RBC suspensions. RBC were spun 7 min at 3100 x g. Resuspension (in buffer A) and centrifugation were repeated until the pellet was hemoglobin-free. The pellet was then washed with low Mg^{2+} (0.2 mM) buffer A and spun 10 min at 3100 x g. The pellet was brought up to 15 ml with low Mg^{2+} buffer A, and homogenized with a drill-driven, tight-fitting pestle (25 passes). The homogenate was sonicated 3 x 10 s with a minicell disruptor, and spun 10 min at 3100 x g. The supernatant was spun 1 h at 100000 x g to yield PM and cytoskeletal elements.

SDS-PAGE electrophoresis was done using 100 μg of the protein dissolved in: 0.17 M Tris/HCl, pH 8.5, 8.7 % SDS, 36 % glycerol, 415 mM β-mercaptoethanol, 0.05 % bromphenol blue. Samples were boiled for 3 min and gradient gels were used to separate the proteins. Gels were stained with coomassie blue, and analysed by densitometry. Fig. 1 shows a representative gel from normoxic and anoxic cells. Comparison of banding patterns showed that anoxia caused a 58.7 $\pm$ 4.0 % (mean $\pm$ se, N = 4, p < 0.01) reduction in the relative amounts of band 2 protein (Table 1).

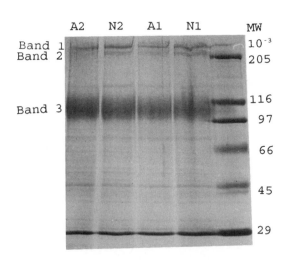

Fig. 1. SDS-PAGE of flounder red cell PM and cytoskeletal elements. N = normoxia and A = anoxia.

The molecular weight of this protein (239 kDa) and relative gel position are consistent with β-spectrin; attempts are underway to confirm this by Western blot analysis. In contrast, anoxia had no effect on other bands, such as band 1 and band 3 (Table 1).

β-spectrin is thought to play a role in maintaining membrane organization and restricting movement of PM proteins of RBC [Bennet, Physiol. Rev. 70:1029, 1990]. Our results suggest that a reduction in β-spectrin under anoxic conditions, with a lessening of the restriction on the mobility of Glc transporters, might explain anoxia-enhanced Glc uptake by flounder RBC. Such a possibility is consistent with the model of Cheung et al. [Biochim. Biophys. Acta 470:212, 1977] to explain anoxia-enhanced Glc transport by avian RBC. From kinetic studies, Cheung et al. postulated that anoxia increased the mobility of Glc transporters by causing loss of bias for its binding site to face in towards the cytosol.

Table 1. High molecular weight proteins of flounder RBC membranes

Band	MW(kDa)	Relative Area		Likely Identity
		Normoxia	Anoxia	
1	255-258	3.1 ± 0.2	3.3 ± 0.7	α-spectrin
2	238-238	1.7 ± 0.2	$0.7 \pm 0.2*$	β-spectrin
3	100-119	24.9 ± 3.4	25.4 ± 2.0	anion exchanger

Mean + SE (4) * $p < 0.005$ vs normoxic control (paired t-test)

Supported by Max-Planck-Society, travel funds to RS and AM-S, Blum-Halsey Award to AM-S and personal monies of GB and RS.

CALCIUM AND MAGNESIUM BINDING IN THE BODY WALL OF THE SEA CUCUMBER CUCUMARIA FRONDOSA

John A. Trotter and Thomas J. Koob
Department of Anatomy, University of New Mexico
Albuquerque, New Mexico 87131

The collagenous tissues of echinoderms have neurally regulated tensile properties. These mutable connective tissues are composed of relatively short (<1mm) tapered collagen fibrils in a nonfibrillar matrix. As is true of any discontinuous fiber reinforced material, the interfibrillar matrix has to transfer stress between the reinforcing collagen fibrils. The stress-transfer properties of the matrix thus control the tensile properties of the tissue, and it follows that neural regulation of tissue compliance is mediated by changes in matrix properties.

The predominant hypothesis for the mechanism by which echinoderms can reversibly change the compliance of their connective tissues is based on divalent ion mediation of collagen fibril interactions. This hypothesis derives exclusivlely from experiments in which artificial perturbation of tissue divalent ion content results in alterations in the tissue's mechanical properties. For example, chelation of divalent ions from the body wall of sea cucumbers greatly increases its compliance (Motokawa, Biol Rev. 59:255, 1984). We here report results of our initial studies that, for the first time, determine the normal ionic composition and divalent ion binding capabilities of the body wall in the sea cucumber Cucumaria frondosa.

C. frondosa in an ideal holothurian for these studies because the body wall lacks ossicles and because there is a conspicuous lack of podia in the ventral interambulacra. These two interambulacral areas were excised and stripped of the inner circular muscle layer. The body wall was then cut into roughly 2 x 4 x 6 mm pieces and the pigmented epidermal layer was removed with a razor blade. For measurement of in situ mineral content, blotted specimens were dried without further manipulation at 65°C. Five specimens from each of 5 animals were analyzed. To determine the proportion of bound mineral, specimens were first thoroughly washed with deionized water and then dried at 65°C. Binding experiments were performed on tissue specimens which had been washed with deionized water, treated with 4 mM ethylenediaminetetraacetic acid in 50 mM Tris-HCl, pH 8.0 to remove bound divalent ions, and again washed with deionized water. Specimens were incubated in artificial sea water (ASW: 10 mM Tris-HCl, 500 mM NaCl, 50 mM $MgCl_2$, 10 mM KCl) containing 0 to 50 mM $CaCl_2$. Following a 24 hr incubation at 12°C, the specimens were thoroughly washed with deionized water and dried. Minerals were eluted from the dried tissue specimens with 1N HCl for 24 hr and quantified by atomic absorption spectroscopy (Ca & Mg) or flame photometry (Na & K). For all the measured values given in Table 1, the coefficient of variation was less than 10%.

The total ion content of the body wall is given in columns 2 and 3 of Table 1. The bound ion content of the water washed tissue is given in column 4 of Table 1. The dry weights following extensive water washing are not comparable to fresh tissue dry weights, since some organic material is eluted in water. Nevertheless, these data indicate that the tissue preferentially

TABLE 1

	Sea Water mmol/L	Fresh Tissue mmol/Kg wet wt.	Fresh Tissue mmol/Kg dry wt.	Washed Tissue mmol/Kg dry wt.	Relative Avidity
Ca	8	8	40	38	4.75
Mg	48	57	285	134	2.79
K	10	24	120	9.5	0.95
Na	500	398	1,975	84	0.17

binds divalent cations, and binds more Ca then Mg. To illustrate this preferential binding, the fifth column in Table 1 gives the ratio of bound ion (mmol/kg dry weight) to the concentration of that ion in sea water (mmol/L). It is seen that the binding avidity is in the order Ca > Mg > K > Na.

The efficiency of mineral elution with increasing pH was determined by incubating specimens sequentially in water, 10^{-6}, 10^{-5}, 10^{-4}, 10^{-3}, 10^{-2}, 10^{-1} and 1 N HCl and measuring the amount of calcium and magnesium that eluted at each concentration. Deionized water removed 63% of tissue magnesium and 38% of tissue calcium. Little additional magnesium or calcium was eluted at pH 6, 5, 4 or 3. The bulk of the remaining mineral was eluted at pH 2. Elution of the last 7 - 10% of the magnesium and calcium required 0.1 N HCl. These results establish that the mineral remaining after water washing is firmly bound to the tissue.

Treatment of water-washed specimens with EDTA removed 98% of bound calcium and magnesium. Calcium binding to these specimens in artificial sea water was concentration dependent between 0.5 and 50 mM. This binding showed complex concentration dependence that could not be attributed solely to single site binding. Nevertheless, it was interesting to note that the amount of calcium bound to tissue incubated in 10 mM $CaCl_2$ (37.4 mmol/kg dry wt.) was nearly identical to that measured in washed fresh tissue (38 mmol/kg).

These results show that calcium binding in the body wall of C. frondosa is both tight and specific. Our binding studies are consistent with the aforementioned mechanical experiments and with our unpublished observations that chelation of calcium is sufficient to produce free collagen fibrils from body wall. The pH at which the bound Ca and Mg were eluted from the tissues corresponds closely to the pK of sulfate groups on glycosaminoglycans (GAGs). This correspondence may be especially meaningful, because sulfated GAGs constitute a major fraction of the interfibrillar matrix of the body wall. Moreover, the principal GAG of the body wall of C. frondosa is closely and periodically associated with the surfaces of the collagen fibrils (Trotter and Koob, Bull. MDIBL 29:28, 1990).

Supported by grants from the NSF and the ONR.

THE ACTION SPECTRUM FOR RELIEF OF CO-INHIBITION IN GASTRIC CELLS FROM RAJA ERINACEA: EVIDENCE FOR AN ALTERNATIVE CYTOCHROME OXIDASE

George W. Kidder III[1] and Brett M. Haltiwanger[2]
[1]Dept. of Biological Sciences, Illinois State University, Normal, IL 61761
[2]Dept. of Biology, Sweet Briar College, Sweet Briar, VA 24595

Acid secretion by chambered skate gastric mucosa drops to near zero in the absence of oxygen. The terminal oxidase inhibitor carbon monoxide (CO) at a high CO/O_2 ratio does not fully block acid secretion (MDIBL Bull. 26:43-46, 1986), nor does it greatly inhibit oxygen uptake by cells from this tissue (MDIBL Bull. 28:14-15, 1989). This small inhibition is partially reversible by light; therefore, the spectrum for the relief of inhibition as a function of wavelength should determine the absorption spectrum of the CO-complex of the functional terminal oxidase. In particular, it should be possible to determine whether the oxidase is the conventional cytochrome oxidase (a_3) or some other pigment. Previous studies (MDIBL Bull. 30:25-26, 1991) suggested a different oxidase, but the spectra were not definitive. Improvements to the apparatus were required for convincing spectra.

Cells were removed from the gastric mucosa by pronase digestion as previously described (ibid.), digesting for 3 hours in 75 ml of Forster's solution containing (mM) NaCl, 200; $NaHCO_3$, 30; KCl, 10; $CaCl_2$, 5; $MgCl_2$, 2; Na_2HPO_4, 1; glucose, 25; urea, 350; plus 140 units/ml pronase (Streptomyces griseus) at 30°C which were washed once in Forster's solution, and stored at 5°C until used. Normally, a preparation made one morning could be used that afternoon and the following day.

Commercial baker's yeast was used as a cytochrome a_3-containing control, using the above solution, with NaCl reduced to 110 mM and no urea, glucose or pronase. A suspension of 10 mg/ml proved satisfactory, with 1% EtOH as respiratory substrate.

The photochemical action spectrum (PCAS) apparatus has previously been described (ibid.). Briefly, we adjusted the intensity of light at variable wavelengths (VW) until it had the same effect on oxygen uptake (oxygen electrode) as a standard intensity of light at a standard wavelength (SW) of 500 nm, and measured the two intensities. The ratio of these intensities is the relative efficiency of light at the VW to that at the SW. The light-measuring system was modified by the addition of a chopper disk which interrups the light beam once per second, and the use of an analog-to-digital converter and an Apple II computer (Comp. Appl. in Biol. Sci. 4:331-335, 1988) to collect, display, average and record the intensity data. Sixteen peak-peak differences were measured, with the mean taken as the intensity; the relative standard error of this determination was about 0.5% at full intensity. The oxygen electrode output was also collected and displayed by the computer, which allowed signal averaging to minimize noise at high sensitivities. With this method, the Teflon membrane could be removed from the Instech electrode, which gave a more rapid response and allowed the use of 2 μl of cell suspension per run. The resulting noise in the oxygen electrode output corresponded to about 0.1% O_2. The gas mixture around the drop was 5% CO_2, 20% O_2 and 75% CO, and the concentration of cells and the position of the electrode was adjusted to give between 1 and 10% O_2 at the electrode tip when illuminated with light at the 500 nm reference wavelength. For each run, a measurement was made with VW = SW = 500 nm. Instrumental differences may make this value not equal unity; therefore, the observed values at other VW's were divided by this correction factor to remove this source of error.

Yeast respiration is sensitive to CO, and this suspension was very responsive in the apparatus, giving the spectrum shown as figure 1, top, which is a good representation of the absorption spectrum of the cytochrome oxidase-CO complex, which agrees with information previously available for cytochrome a_3. Skate cells were less responsive to light, with some preparations being completely insensitive, which introduces errors into the determination and results in a noisy spectrum, as shown in figure 1, bottom. The most effective wavelength is in the vicinity of 580 nm in skate, while it is clearly at 590 nm in yeast. The ratio of spectral height at 590 to that at 580 was 0.75 in skate cells, while it was 1.57 in yeast cells.

While another peak is expected near 430 nm, the performance of the apparatus in this wavelength range did not give confidence in the results. Due to the characteristics of the lamp and monochromators, the maximum intensity which could be generated at 400 nm is only 1/10 that available at 500 nm, which seems to be too little for accurate results.

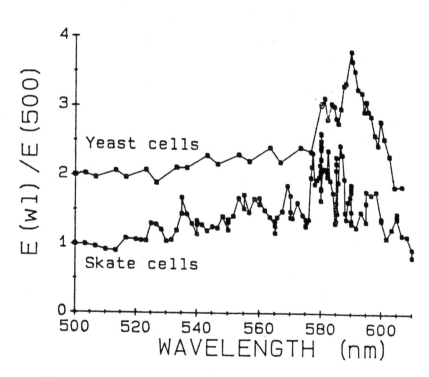

Figure 1:
Action spectra for the relief of CO-inhibition in yeast and skate cells. Data (yeast N=57, skate N=148) smoothed by a 3-point formula, with connecting lines. Yeast spectrum displaced +1 unit.

The ability of this tissue to continue to secrete in the presence of high $CO-O_2$ ratios, the insensitivity of its oxygen uptake to CO, N_3^- and CN^-, the appearance of a reduced cytochrome peak at 590 nm in the presence of N_3^-, and now the identification of a ca. 580 peak of the CO-complex of the functional oxidase all suggest that a component of the oxygen uptake in this tissue is mediated by an oxidase which is not the conventional mitochondrial cytochrome $a+_3$. While this oxidase remains to be isolated and characterized, it may well be localized to the plama membrane and play a role in the secretion of H^+ by this tissue, as previously suggested (Ann. N.Y. Acad. Sci. 574:219, 1989).

(Supported in part by a grant from the Pew Memorial trust to BMH)

PREPARATION AND CHARACTERIZATION OF IMMOBILIZED ANTIPAIN AND CHYMOSTATIN FOR THE ISOLATION OF PROTEOLYTIC ENZYMES BY AFFINITY CHROMATOGRAPHY

J. William Straus[1], Catherine L. Monian[1], and David L. Cox[2]
[1]Department of Biology, Vassar College, Poughkeepsie, NY 12601
[2]Department of Biology, University of Oregon, Eugene, OR 97403

Proteolytic enzymes are nearly ubiquitous in cells as well as in extracellular fluids and structures. Our interest in the isolation of proteinases stems from studies of the production and secretion of proteinases in relation to cellular growth and development in ciliate protozoans, specifically Tetrahymena thermophila. (Straus, J. Protozool. **37**, 9A, 1990). While affinity chromatography with immobilized inhibitors is a well established technique for the isolation of proteases, there has been very little use of the peptide aldehyde inhibitors (Umezawa, Meth. Enz. **45**, 678-695, 1990). Peptide aldehydes such as leupeptin, calpain inhibitors I & II, antipain, and chymostatin are exceptionally strong inhibitors of proteolytic activity in cell lysates of T. thermophila (Straus, 1990, op cit.). Moreover, antipain and chymostatin each have a free carboxyl group distal to the inhibitory aldehyde moiety, which makes them suitable for linkage to an immobile amine group via a condensation reaction. Hence, we prepared antipain-agarose and chymostatin-agarose and began to characterize their chromatographic properties as part of an ongoing effort to rapidly isolate proteolytic enzymes from cellular lysates and secretions.

Antipain, chymostatin, and chymotrypsin were purchased from Sigma Chemical Company and papain was obtained from Boehringer Mannheim Biochemicals. Cellular lysates of T. thermophila (strain WH-14) were prepared from axenic cultures harvested at early stationary phase (Straus, 1990, op cit.). Diaminodipropylamine-agarose (DPA) and ethyldimethyl-aminopropyl carbodiimide (EDC) were purchased from Pierce Chemical Co. For inhibitor immobilization, 5 mg aliquots of antipain or chymostatin in 1 ml dimethylsulfoxide were mixed with 1 ml DPA in 50 mM acetate at pH 4.8, and coupled via an amide linkage by the addition 50 mg EDC (Chase, Merrill, & Williams, Proc, Natl. Acad. Sci. USA **80**, 5480-5484, 1983). Coupling reactions were incubated in the dark with slow shaking for 12 h. Following coupling, the immobilized matrix was placed in a 5 ml chromatography column and washed with at least 50 volumes of 0.5 M NaCl, 10 mM HEPES, pH 7. For storage, columns were equilibrated with wash buffer containing 0.02% NaN_3 at 4° C. Control columns were prepared by incubating DPA with EDC in the absence of peptides. Prior to loading proteinases on columns, papain and chymotrypsin were dissolved in wash buffer at concentrations of 1 mg/ml and T. thermophila lysates were diluted with 0.1 volume of 10X wash buffer.

Aliquots (0.5 ml) of papain or chymotrypsin (1 mg/ml) were applied to columns (1 ml bed) and eluted with at least six volumes wash buffer. With papain and T. thermophila lysates, 1 mM dithiothreitol was included in the initial wash buffer to stabilize active site cysteines and to promote enzyme-inhibitor interactions. Subsequent elutions were performed with acidic buffer (0.05 M citrate, 0.5 M NaCl, pH 3) and chaotropic agents (6 M guanidine-HCl or 8 M urea in wash buffer). Column eluants were monitored spectrophotometrically at 280 nm to estimate protein concentrations. Proteolytic activity in column fractions was assessed by following the digestion of azocasein spectrophotometrically at 440 nm (Straus, Parrish, & Polakoski, J. Biol. Chem. 256, 1981).

When 0.5 ml aliquots of papain (1 mg/ml) were applied to a control column, all of the protein and enzyme activity eluted in the first three ml, with 90% of the protein and activity in the 2 ml fraction. No additional protein or proteolytic activity was eluted when the column was washed with acidic buffer or 6 M guanidine-HCl (Fig. 1A). In contrast, when applied to an antipain-DPA column, only about 10% of the protein and less than 1% of the initial activity were recovered in the 2 ml fraction. Secondary washing with acidic buffer eluted less than 25% of the total protease activity and about 22% of the total protein. Further washing with guanidine-HCl eluted the bulk of protein, but no enzymatic activity was recovered in those fractions (Fig. 1B). Elution of protein and enzymatic activity by acidic buffer on subsequent trials was typically less than shown below. A small amount of protein without activity was recovered when the column was eluted with 8M urea (Fig. 1C).

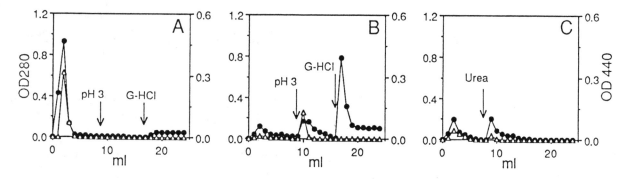

Fig. 1. Affinity chromatography of papain with antipain-DPA. OD 280 (●) indicates spectrophotometric absorption at 280 nm while OD 440 (Δ) indicates spectrophotometric absorption at 440 nm of azocasein digests, prepared by incubating 50 µl aliquots from each fraction with 0.1% azocasein in 10 mM HEPES, pH 7 for 1 h at 35° C. A, elution of papain from control DPA column. B, elution of papain from antipain-DPA column with pH 3 buffer and 6 M guanidine-HCl (G-HCl) at pH 7. C, elution of papain from antipain-DPA column with 8 M urea at pH 7.

Chymotrypsin behaved similarly to papain when applied to the control column, as shown in Fig 2A. There was no apparent retention of protein or enzymatic activity by the DAB matrix and further washing with acidic buffer and 6 M guanidine-HCl caused no apparent release of protein or activity. When applied to the chymostatin-DPA column, over 95% of the protein and all of the detectable enzymatic activity were retained (Fig 2B & 2C). Washing with pH 3 buffer caused a very slight release of active enzyme (less than 10% of the total applied activity) (Fig. 2B). Guanidine-HCl and urea both caused gradual elution of protein and proteolytic activity and in both cases the enzyme was spread over a large number of fractions (Fig. 2B & 2C).

When T. thermophila lysate was applied to the DPA column, there was no detectable retention of either protein or proteolytic activity (Fig. 3A) and as with papain and chymotrypsin, further washing with acidic buffer and guanidine-HCl caused no apparent release of protein or enzyme activity. Immobilized antipain retained 65% of the total applied protein and over 90% of the proteolytic activity (Fig. 3B). No activity or protein was eluted by acid buffer. The bulk of applied protein was eluted by 6 M guanidine-HCl, but no activity was detected in these fractions (Fig. 3B). Similar results were obtained when cell lysates were applied to chymostatin-DPA (not shown). Over 50% of lysate protein and over 80% of lysate proteolytic activity were retained by chymostatin-DPA (Fig. 3C). Application of 6 M urea caused a very slight and gradual release of protein and enzyme activity. Urea caused a slight release of protein without activity from antipain-DPA (not shown).

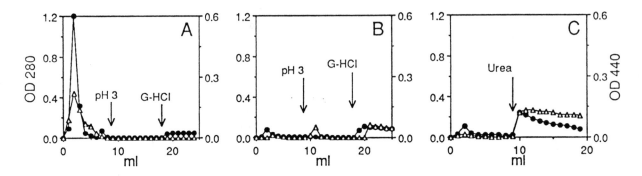

Fig. 2. Affinity chromatography of chymotrypsin with chymostatin-DPA. OD 280 (●) and OD 440 (Δ) are as described in Fig. 1. **A**, elution of chymotrypsin from control DPA column. **B**, elution of chymotrypsin from chymostatin-DPA column with pH 3 buffer and 6 M guanidine-HCl (G-HCl) at pH 7. **C**, elution of chymotrypsin from chymostatin-DPA column with 8 M urea at pH 7.

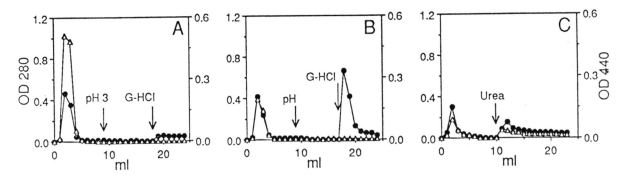

Fig. 3. Affinity chromatography of T. thermophila lysate proteases with antipain-DPA and chymostatin-DPA. OD 280 (●) and OD 440 (Δ) are as described in Fig. 1. A 1.0 ml aliquot of cell lysate was applied to columns 3A and 3C and a 2 ml aliquot was applied to column 3B. The lysate protein concentration was estimated to be 0.6 mg/ml based on absorbance at 280 nm against a bovine serum albumin standard. **A**, elution of lysate proteases from control DPA column. **B**, elution of lysate proteases from antipain-DPA column with pH 3 buffer and 6 M guanidine-HCl (G-HCl) at pH 7. **C**, elution of lysate proteases from chymostatin-DPA column with 8 M urea at pH 7.

The above results indicate that both antipain and chymostatin were successfully coupled to DPA and that both affinity matrices retained proteolytic enzymes. Attempts to recover bound protein and proteolytic activity by elution at pH 3 met with very limited success. A small proportion of bound papain and cell lysate protease were recovered when the antipain-DPA column was eluted with acidic buffer (Figs. 1B & 3B) while no protein or activity was recovered from the chymostatin-DPA column under similar circumstances (Fig. 2B). Attempts to remove bound enzymes by denaturation with chaotropic agents also met with limited success. A substantial amount of protein, but no enzymatic activity was recovered from the antipain-DPA column when washed with 6 M guanidine-HCl (Figs. 1B & 3B). Active chymotrypsin was recovered from the chymostatin-DPA column when eluted with 6 M guanidine-HCl (Fig. 2B) or 8 M urea (Fig. 2C), but release of the enzyme was slow and spread over many fractions. Urea had similar, but less dramatic effects on the elution of protein and enzyme from antipain-DPA (Fig. 1C). While the immobilized inhibitors were very effective at retaining proteolytic enzymes, acidic and chaotropic conditions were relatively ineffective in eluting active enzyme from the affinity matrices. Accordingly, more effective elution conditions will be needed to if these immobile matrices are to be used for enzyme purification. Other eluants to be tested will include active site ligands (soluble inhibitors or substrates) to compete with the immobilized ligands for enzyme.

(Supported by fellowships to J.W.S. and C.L.M. from the Pew Charitable Trust and by a William and Flora Hewlett Foundation Grant of Research Corporation to J.W.S.)

CATECHOL OXIDASE LATENCY IN NIDAMENTAL GLAND EXTRACTS FROM THE LITTLE SKATE (RAJA ERINACEA)

J. William Straus[1], Catherine L. Monian[1] and David L. Cox[2]
[1]Department of Biology, Vassar College, Poughkeepsie, NY 12601
[2]Department of Biology, University of Oregon, Eugene, OR 97403

Egg capsules of the little skate (Raja erinacea) are formed from protein precursors secreted by the nidamental gland. Sclerotization of the capsule involves the production of catechols and their subsequent oxidation to quinones by catechol oxidase. This enzyme is latent when incorporated into the capsular matrix, and is activated coordinately with increasing capsule catechol content and color development (Koob & Cox, J. mar. biol. Assoc. U.K. 70, 395-411, 1990). In nidamental gland lysates, catechol oxidase activity is typically preceded by a latent period of one to several minutes following substrate addition. Treatment of catechol oxidase with α-chymotrypsin significantly reduced the latent period and concomitantly increased initial rates of substrate oxidation. Conversely, treatment of nidamental gland extracts with a serine proteinase inhibitor caused substantial increases in oxidase latency and reduction in activity. Thus, a proteinase or proteinase sensitive factor might influence catechol oxidase activity in vivo (Koob & Cox, Biol. Bull. 175, 202-211, 1988: Cox & Koob, Comp. Biochem. Physiol. 95B, 767-771, 1990). Subsequent investigations of enzyme latency revealed that nidamental gland lysates contain an endogenous factor which promotes latency without affecting the initial rate of substrate oxidation (Straus & Cox, Bull. Mt. Desert Is. Biol. Lab. 30, 23-24, 1991). The latency factor was shown to extend the latent period in a concentration dependent manner. Further characterization revealed that the factor was insensitive to heat denaturation, destroyed by acid hydrolysis, and apparently was of low molecular weight. Our continued studies have focused on the effects of substrate, ions and chelation on catechol oxidase latency, and on the molecular properties of the latency factor.

Nidamental gland lysates were prepared by homogenization as previously described (Koob & Cox, 1988, op cit.) in 0.5 M NaCl buffered at pH 7 with 50 mM Tris-HCl or 10 mM HEPES. Glands were divided into equivalent sections for homogenization in Tris extraction buffer in the presence or absence of 25 mM ethylenediaminetetraacetic acid (EDTA). Five ml aliquots of the control and EDTA extracts were dialyzed against extraction buffer (to a dilution factor of 4 X 10^{11}) at 4° C with Spectrapor 2 membrane to remove small molecules and chelating agent. Enzyme activity was determined by following the oxidation of 4-methylcatechol at 400 nm as previously described (Koob & Cox, 1988, op cit.).

Extracts prepared with EDTA exhibited a six-fold increase in the latent period following substrate addition, and they oxidized substrate (1 mM 4-methylcatechol) at rates 6.5% (N=12) of control extracts. It should be noted that dialysis of EDTA extracts restored catechol oxidase activity to the same levels as dialyzed control extracts. Dialysis eliminated the initial latent period in both control and EDTA extracts, and increased oxidation rates 2.2- and 2.7-fold respectively over those of undialyzed control extracts.

Pre-incubation of control extracts with physiological concentrations of $CaCl_2$ (6 mM) or $MgCl_2$ (2.5 mM) for 15 min at ambient temperature lengthened the initial latent phase by about three-fold and diminished oxidase activity by 50 - 60%. EDTA extracts were similarly affected by treatment with CaCl$_2$ and unaffected by treatment with MgCl$_2$. Pre-incubation of dialyzed and control extracts with MgCl$_2$ had no apparent effect on either latency or activity. Conversely, pre-incubation of dialyzed extracts with CaCl$_2$ for 15 min restored the latent period which dialysis had abolished and reduced oxidation rates of both extracts by over 90%. It is also important to note that when lysates from other animals were examined, the effects of chelation and divalent cations varied considerably and in one case, chelation and divalent cations had virtually no effect on catechol oxidase activity in the nidamental gland lysate.

The absolute period of oxidase latency tended to be fairly consistent when right and left nidamental gland extracts from the same animals were compared. Intraorganismal variation was typically on the order of ±1 min or less. However, interorganismal differences in latent periods varied from about 1.5 to 12 min. We also found that the concentration of 4-methylcatechol substrate used in assays had a dramatic impact on enzyme latency, as shown in Fig. 1. Increasing substrate concentration caused a corresponding rise in oxidase activity with a concomitant decrease in latency. A ten-fold increase in substrate concentration resulted in a 60% decrease in the latent period. Further increases in substrate caused diminution of latency in smaller increments.

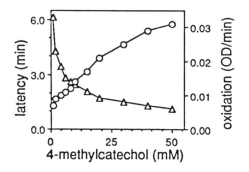 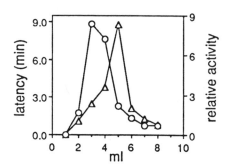

Fig. 1 (left) - Effect of substrate concentration on catechol oxidase latency (Δ) and activity (O). Aliquots of lysate (0.05 ml) were equilibrated for 2 min in 10 mM HEPES, 0.5 M NaCl, pH 7 at ambient temperature. Reactions were initiated by the addition of substrate.

Fig. 2 (right) - Gel filtration of nidamental gland lysate. The column contained a 2 X 0.5 cm bed of Sephadex G-25. 0.5 ml of lysate with an endogenous latency of 12.1 ± 1.2 min was applied and eluted with 10 mM HEPES, 0.5 M NaCl, pH 7. 1 ml fractions were collected and assayed for oxidase activity as above. Latent period (Δ) and relative activity (O) for each fraction are shown, where one unit of relative activity equals 0.001 O.D. units/min at 400 nm.

Extracts prepared in 10 mM HEPES responded to dialysis in a similar manner to those above. For example, dialysis to a dilution factor of 4000 resulted in a 70% decrease in latency (from 2.2 ± 0.2 min to 0.7 ± 0.1 min) while a dilution factor of 10^6 resulted in complete loss of the latent period. In addition, dialysis caused a doubling in the initial rate of substrate oxidation. Because latency was reduced or eliminated by dialysis, extracts were subjected to gel filtration to determine whether the latency factor could be separated from catechol oxidase. Prior results have shown that when an aliquot of nidamental gland lysate was centrifuged through a bed of Sephadex G-25, a reduction in latency resulted (Straus & Cox, 1991, op cit.). A short (2 cm) Sephadex G-25 column was used to minimize dilution and separation time for chromatographic fractionation. As shown in Fig. 2, when an extract was subjected to gel filtration, the oxidase and the latency factor separated into two distinct but overlapping peaks. The first three fractions containing oxidase activity (collected at 2, 3, and 4 ml respectively) exhibited highly reduced latency while the subsequent fraction (collected at 5 ml) retained a nine minute latent period.

Glutathione, a thiol reducing reagent, proved to be a potent mimic of the endogenous latency factor. As shown in Fig. 3A, addition of reduced glutathione to nidamental gland lysate resulted in a concentration dependent increase in the latent period. Reduced glutathione had no apparent effect on initial rates of substrate oxidation. In contrast, oxidized glutathione had no effect on either latency or activity (Fig. 3B). Another thiol reductant, dithiothreitol, also increased latency in a similar manner without affecting activity.

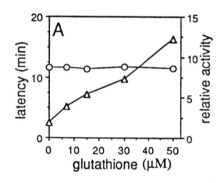 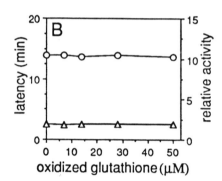

Fig. 3. Effect of reduced (**A**) and oxidized (**B**) glutathione on catechol oxidase latency (△) and activity (○). Aliquots of extract were equilibrated for 2 min with the indicated concentrations of glutathione at ambient temperature. Reactions were initiated by the addition of substrate and monitored for oxidase activity. For relative activity see Fig. 2.

The regulation of catechol oxidase is essential for egg capsule formation and hence, reproduction in the little skate. The results presented above provide intriguing information on the behavior of nidamental catechol oxidase. We have established that lysates contain an endogenous factor which induces temporary enzyme latency in a concentration dependent manner (Straus & Cox, 1991, op cit.). The factor is resistant to heat denaturation, destroyed by acid hydrolysis, and removable by dialysis and gel filtration. These data indicate that the latency factor has a low molecular weight (< 12, 000) and suggest that the factor could be a peptide since it was susceptible to acid hydrolysis. The ability of glutathione and dithiothreitol to induce temporary latency in a manner similar to the endogenous factor suggest that the factor might function as an antioxidant. The effects of chelation and divalent cations are difficult to interpret since there was extensive interorganismal variation. Clearly, metals have profound effects on catechol oxidase latency and activity in some, but not all nidamental gland lysates. The ability of calcium to dramatically increase latency and diminish oxidase activity in certain extracts indicates that enzyme sensitivity to divalent cations might depend on the stage of the reproductive cycle. Similarly, the considerable interorganismal variation in catechol oxidase latency and activity might arise from reproductive stage differences. A thorough understanding of this system will require biochemical characterization of purified catechol oxidase and regulatory components, as well as an examination of the physiological parameters and reproductive stage effects pertinent to enzyme secretion and activation.

(Supported in part by fellowships to J.W.S. and C.L.M. from the Pew Charitable Trust)

EGG CASE PERMEABILITY TO AMMONIA AND UREA
IN TWO SPECIES OF SKATES (RAJA sp.)

Gregg A. Kormanik[1], Antoine Lofton[1] and Niamh O'Leary-Liu[2]

[1]Department of Biology, Univ. of North Carolina at Asheville, NC 28804
[2]P.O. Box 218, Surrey, ME 04684.

Embryos of oviparous skates are enclosed in egg capsules and bathed by sea water for a gestation period which may last nearly two years. While the tough capsule mechanically protects the embryo, its role as an osmotic barrier has been less clear. Smith (Biol. Rev. 11:49-82, 1936) proposed that the "cleidoic" egg of elasmobranchs was an adaptation for retention of urea. An assumedly impermeable egg capsule is cited by several authors in recent discussions of the evolution of viviparity and urea retention (see Kormanik, Am. Zool., in press). To clarify the role of the egg capsule in osmoregulation, we measured the permeability of the capsule wall to ammonia and urea.

Female skates (Raja erinacea and one other species, see below) were collected by commercial fishermen from Frenchman Bay, ME., and maintained in running sea water. Fresh eggs in capsules were collected daily, and used in subsequent experiments. For the measurements of permeability, sections (2 cm^2) were removed from the capsule, mounted in a lucite Ussing-type chamber, filled with the appropriate solution and bubbled with air. Artificial sea water (Na^+, 450; Cl^-, 525; Ca^{++}, 10; Mg^{++}, 56; K^+, 9; SO_4^{--}, 32; HCO_3^-, 2.5 mM; buffered with TRIS, 5 mM, in the pH experiments) was placed on the outside and inside. A stock solution (1 M) of either urea or ammonia was added to the inside to achieve final concentrations of 100 mM and 10 mM respectively. The sea water concentration was previously adjusted to minimize any osmotic gradient across the capsule wall. In a second series of measurements of ammonia diffusion, the pH of the ASW was adjusted to modify separately the gradients for NH_3 and NH_4^+ (calculated after Cameron and Heisler, J. Exp. Biol. 105;107-125, 1983). The appearance of urea or ammonia on the outside of the capsule wall was measured at 30 min intervals using the diacetyl monoxime method (Sigma Kit # 535) for urea or the Solorozano method (Limnol. Oceanogr. 14;799-801, 1969) for ammonia. The permeability (Px) was calculated from the equation $J_{net} = P_x * A * \Delta C$ where J_{net} is the net flux, A is the area and ΔC, the concentration gradient. During the course of the experiment (2 hrs.) the gradients changed less than 5%, but were nevertheless corrected for; the net fluxes were linear. Results are expressed as mean ± 1 S.E.M.

The results of the first series of experiments are presented in Table 1. The permeability we measured for urea, while lower, is comparable to that measured by Hornsey for Scyliorhinus canicula (4.7×10^{-5} cm sec^{-1}; Experientia 34;1597, 1978). The egg case is also quite permeable to ammonia.

Elevating pH had only a small effect on the NH_4^+ gradient while the NH_3 gradient was enhanced 20-fold (Table 2). There was no significant difference in total ammonia diffusion. Either ammonia diffuses predominantly as the charged form, NH_4^+, or the capsule wall does not distinguish between these two species. Permeabilities calculated from these data (from Table 2 using two equations for two unknowns) were 2.2 and 2.5 cm sec^{-1}, for NH_3 and NH_4^+, respectively. Hornsey (ibid.) characterized the capsule wall of S. canicula as having pore radii on the order of 1.4 nm,

Table 1. Permeability of the egg capsule wall of skates to ammonia and urea. T_{amm} is total ammonia. Permeability (P_x) is expressed in cm sec^{-1}, all values x 10^5, n = number of capsules.

	P_{urea}	P_{Tamm}	Thickness (cm)
R. erinacea (8)	2.00 ±0.15	3.09 ±0.16	0.033 ±0.002
R. ocellata* (1)	1.30	----	0.066

* -mother unidentified, tentative classification is based on egg morphology (Bigelow and Schroeder, Fishes of the Western North Atlantic, Memoire I, Pt. II., p.247, Sears Fnd. Mar. Res., New Haven, 1953)

Table 2. Total ammonia fluxes across the egg capsule of R. erinacea when the gradients for NH_4^+ and NH_3 are independently varied. Outside pH was always 7.0, inside pH is given below. Gradients are "inside" - "outside", (n = 5-6).

Expt.	ΔNH_4^+ (mM)	ΔNH_3 (mM)	J_{net} (x 10^{-10} mol cm^{-2} sec^{-1})
Control (pH = 7.0)	9.97 ± 0	0.021 ± 0.001	2.49 ± 0.65
Exptl. (pH = 8.3)	9.58 ± 0.04	0.42 ± 0.04	2.52 ± 0.34
(Exptl/cntrl)	0.96	20.	1.01
Signif. (p<)	0.001	0.001	n.s. (p>0.2)

much larger than the hydrated radii for both of these highly water soluble forms, as well as urea and water molecules. The egg capsule wall is quite unlike elasmobranch branchial membranes, where P_{NH3}/P_{NH4}^+ is quite large (Evans & More J. Exp. Biol. 138:375-397, 1988).

These data confirm our previous observations, and help to explain why neither ammonia nor urea accumulates in capsular fluids (Kormanik, Bull. MDIBL 28;12-13, 1989). Since large gradients exist for the nitrogenous compounds urea and TMAO across the egg membrane (see Kormanik et al., this issue), embryonic membranes appear to be the main barriers to diffusion and not the capsule wall. While only a few oviparous species have been examined, all have shown high capsular permeabilities to small molecules and ions. The relatively high permeability of the capsule should be recognized in discussions of the evolution of urea (and TMAO) retention in the elasmobranchs (Supported by NSF DCB-8904429 to GAK and and a Hearst Foundation, Inc. scholarship to NOL.)

TYROSINE HYDROXYLATION DURING EGG CAPSULE TANNING
IN THE LITTLE SKATE RAJA ERINACEA

Thomas J. Koob
Mount Desert Island Biological Laboratory
Salsbury Cove, Maine 04672

Six major proteins secreted and assembled by the nidamental gland form the egg capsule of the little skate (Koob & Cox, Bull. MDIBL 30, 21-22, 1991). Sclerotization of these capsule precursors is accomplished by a form of quinone tanning in which catechols are introduced into the capsular matrix following secretion and are then oxidized to quinones by catechol oxidase (Koob & Cox, J. mar. biol. Assoc. U.K. 70, 395-411, 1990). The present report identifies the origin and nature of the introduced catechols.

Egg capsules at various stages of formation were removed from the uterus for tyrosine hydroxylase assays and for quantitation of tyrosine and 3,4-dihydroxyphenylalanine (DOPA). The newly secreted portion of partially formed capsules is untanned and chemically unstable, it contains no catechol, and the catechol oxidase is inactive (Koob & Cox, op. cit.). Quinone tanning progresses with time after formation in the nidamental gland so that older portions of partially formed capsules have begun to tan (see Koob & Cox, Bull. MDIBL 26, 109-112, 1986). Fully formed but partially tanned capsules also show a gradient of capsule tanning. Fully tanned egg capsules removed from in utero as well as capsules collected at oviposition were likewise analyzed for tyrosine hydroxylase activity and tyrosine/dopa contents.

Body walls of capsules were dissected into six sequential 1 x 2 cm specimens cut perpendicular to the long axis of the capsule. Each specimen was dissected into four equivalent pieces as replicates for analyses. Tyrosine hydroxylase activity in the capsular material was measured directly by incubating specimens with 1 mM tyrosine in 3 ml of 0.1 M NaH_2PO_4, 0.45 M NaCl, pH 7.0 containing 25 mM ascorbate to prevent oxidation of DOPA. At 30 minute intervals 50 μl samples of the reaction mixture were collected for quantitation of tyrosine and DOPA by reverse phase HPLC according to Marumo & Waite (Biochim. Biophys. Acta 872: 98-103, 1986). Rates of tyrosine hydroxylation by capsule specimens were routinely compared to boiled control specimens incubated in parallel. To determine whether tyrosine hydroxylation occurs naturally during the tanning process, tyrosine and DOPA contents in eight sequential specimens from partially formed, partially tanned and fully tanned capsules were measured by reverse phase HPLC after hydrolysis of lyophilized specimens in 6 N HCl, 108°C for 24 hr.

Initial experiments verified that a tyrosine hydroxylating activity was present in the capsule matrix: capsule specimens catalyzed the hydroxylation of tyrosine forming DOPA and the rate of tyrosine hydroxylation was linear over the 3 hr assay period; there was equimolar stoichiometry between tyrosine lost and DOPA produced in the tyrosine hydroxylase assays; hydroxylating activity was sensitive to the pH of the reaction mixture; boiled capsule specimens failed to hydroxylate tyrosine thus indicating the enzymatic nature of the hydroxylating activity.

The effect of substituted tyrosine analogues on tyrosine hydroxylation by egg capsule specimens was determined by quantifying hydroxylation rates in the presence of each analogue at a concentration equimolar with tyrosine (1 mM). The three analogues tested were α-methyl-L-p-tyrosine (metyrosine: a specific tyrosine hydroxylase inhibitor), α-methyl-DL-m-tyrosine and p-amino-L-phenylalanine. Metyrosine was hydroxylated at rates nearly identical to those of tyrosine indicating that substitution on the α-carbon of the amino acid does not prohibit ring hydroxylation. α-Methyl-DL-m-tyrosine and p-amino-L-phenylalanine were not hydroxylated nor did they interfere with tyrosine hydroxylation, indicating that the position and nature of the ring substitution are critical for enzymatic activity. These results showed that the capsule hydroxylase required a ring para-hydroxyl group. Together these results indicate that the substrate specificity of the capsule hydroxylase relies on the hydroxyphenol side chain of the amino acid residue. Since metyrosine failed to inhibit hydroxylase activity, the enzyme appears different from the tyrosine hydroxylase involved in catecholamine synthesis.

Measurement of tyrosine hydroxylase activity in six sequential specimens from a partially formed egg capsule showed site-related variations in enzymatic activity that correlated with the tanning process (Table 1A). Lowest hydroxylase activity was found in the pretanned newly secreted capsular material still within the gland lumen. Hydroxylase activity increased with time after secretion reaching 3-fold greater activity in the older tanning specimens. This pattern correlated with the disappearence of tyrosine and formation of DOPA during the tanning process in situ (Table 1B). Specimens from the pretanned portion of a partially formed capsule contained no detectable DOPA. In sequential specimens DOPA contents increased and tyrosine contents coordinately decreased with time after secretion. These results indicate that formation of DOPA via tyrosine hydroxylation occurs naturally during the tanning process.

TABLE 1. A. Tyrosine hydroxylase activity in sequential specimens from partially formed egg capsule.

	pretanned				tanning	
	1	2	3	4	5	6
μmoles/min	0.7	0.8	1.3	1.8	2.2	1.9

B. Tyrosine and dopa contents in sequential specimens from partially formed egg capsule.

	pretanned							tanning
μmoles/mg	1	2	3	4	5	6	7	8
tyrosine	1.22	1.20	1.17	1.16	1.07	0.98	0.84	0.70
dopa	0	0	0.01	0.04	0.08	0.13	0.18	0.27

Tyrosine levels in capsules further along in the tanning process were lower than those of partially formed capsules while DOPA contents were higher (Table 2). Lowest tyrosine levels and highest DOPA contents were found in fully formed capsules from in utero and capsules at oviposition. Overall during the tanning process, tyrosine levels declined by about 50% while DOPA contents went from zero in the pretanned capsule to levels nearly equimolar with tyrosine in fully tanned specimens. The apparent hydroxylation of tyrosine occurring during the tanning process is consistent with the relatively constant levels of measured tyrosine hydroxylase activity in capsules at these same stages of tanning (Table 2).

Table 2. Tyrosine hydroxylase activity, tyrosine and dopa contents in partially formed/ partially tanned (PF/PT), fully formed/partially tanned (FF/PT), fully formed/ fully tanned (FF/FT) capsules removed from in utero and in capsules at oviposition (AO).

	PF/PT	FF/PT	FF/FT	AO
hydroxylation (μmoles/min)	1.4	1.6	2.3	1.3
tyrosine (μmoles/mg)	1.04	0.85	0.49	0.50
dopa (μmoles/mg)	0.09	0.15	0.16	0.17

These results establish that tyrosine hydroxylating activity is instrumental in the quinone tanning process that operates during egg capsule formation in the little skate. Tyrosine residues present in egg capsule precursors are hydroxylated by this enzyme forming DOPA which is later oxidized to dopachrome by catechol oxidase. It seems likely that the native substrate for this hydroxylase activity is tyrosine contained in one or more of the tyrosine-rich capsule proteins identified earlier (Koob & Cox, op. cit.). The tyrosine hydroxylase could act autonomously, or alternatively, it could be structurally related to the catechol oxidase, i.e. as part of a tyrosinase. In either case, the presence of enzymes typical of melanizing tissues and of melanin-like substances in the tanned egg capsule (Koob, J. mar. Biol. Ecol. 113, 155-166, 1987) suggests that the nidamental gland cells which produce these enzymes develop from neural crest derivatives.

Funded by the investigator

Disruption of Protein Metabolism in Squalus Acanthias Spermatocysts by Mercurials

David M. Barnes and David S. Miller
Laboratory of Cellular an Molecular Pharmacology, NIEHS-NIH
Research Triangle Park, N.C. 27709

One consequence of heavy metal exposure is impaired reproduction (Clarkson et al, Reproductive and Developmental Toxicity of metals, Plenum Press, NY, 1983). Most studies in this area have focused on effects on the female reproductive tract, thus little information is available concerning the effects of heavy metals on spermatogenesis. Dogfish (Squalus acanthias) testis provides a unique model in which to study spermatogenesis. Unlike mammals, the dogfish testes are separated into distinct zones of germ cell maturation (Zones I, II, and III, or premeiotic, meiotic, and postmeiotic stages of spermatogenesis) and spermatocysts from each zone can be isolated and maintained in long-term culture (Callard et al. J. Exp. Zool. Suppl., 2:23, 1989). Described here are initial experiments in which we examined the effects of two mercurials (mercuric chloride, $HgCl_2$ and p-chloromercuriphenyl sulfonic acid, pCMBS), on ^3H-leucine labelling of the intracellular amino acid pool and protein in spermatocysts from the three zones.

Spermatocysts were isolated and cultured by the methods of Callard and Dubois (Bull. MDIBL 27:30, 1988). For experiments, cysts were incubated at 20°C and exposed to mercury for 24 hours; ^3H-leucine was added to the medium for last hour of exposure. Cysts were separated from the media by centrifugation, washed three times and treated with 10% trichloroacetic acid (TCA). Preliminary studies showed that TCA precipitable label was protein bound since 10 μM cycloheximide blocked >95% of the incorporation of label into this compartment. Label in the supernatant was derived from the intracellular amino acid pool and, as expected, was not affected by cycloheximide.

Figure 1 shows the results of 24 hours of $HgCl_2$ exposure on ^3H-leucine labelling of the two intracellular amino acid compartments. In spermatocysts from all three zones, mercury at 30 μM or below had no inhibitory effect on incorporation into protein, but between 30 and 100 μM, incorporation was essentially abolished. In Zone III spermatocysts, 1-30 μM $HgCl_2$ significantly stimulated incorporation (P<0.05 vs controls). A different pattern of effects was observed on labelling of the intracellular amino acid pool. Label in this pool was reduced significantly in Zone I after exposure to $HgCl_2$ concentrations as low as 1 μM; Zone II showed a significant decrease at 10 μM and greater; and Zone III was unaffected below 100 μM.

In similar experiments with pCMBS, a poorly permeable and less potent mercurial, we found that labeling of the intracellular amino acid pool and cell protein fell in parallel in all zones (Fig. 1). Thus, for pCMBS, the effects on protein synthesis could be explained primarily by a reduction in intracellular leucine label, presumably caused by inhibition of leucine uptake at the plasma membrane. In contrast, because a substantial fraction of $HgCl_2$ is not ionized in aqueous solutions, this mercurial is expected to rapidly penetrate cell membranes and act at both plasma membrane and intracellular sites. Consistent with intracellular actions Fig. 1 shows that in all three zones 100 μM $HgCl_2$ reduced leucine incorporation into protein to a much greater extent than labelling of the free amino acid pool. However, Fig. 1 also shows that in Zone III low concentrations of HgCl$_2$ stimulate labelling of protein, but did not affect labeling of the amino acid pool. In Zone I, those same low concentrations reduced labeling of the amino acid pool but had no effect on the labeling of protein. One explanation for these findings is that HgCl$_2$ had a biphasic effect on absolute rates of protein synthesis in Zones I and III not occurring in Zone II. This biphasic effect was manifest by stimulation at low concentration and inhibition at higher concentrations.

Together, the present data indicate that protein metabolism in dogfish spermatocysts is disrupted by HgCl$_2$ and that some of the effects may be stage specific. Spermatocysts contain two interacting cell types, germ cells and sertoli cells. During spermatogenesis, changes take place not only in the relative proportion of the two cells within the cyst, but also the biochemical processes occurring within each. Thus it is important that we understand at each stage of spermatogenesis how toxins act in terms of their effects on each cell type and on the interaction between cells. This is a goal of future studies using the dogfish model.

FIGURE 1. Effects of mercurials on labeling of the intracellular free amino acid (AA) and protein pools by ^{3}H-leucine. Spermatocysts from Zones I, II, and III were exposed to the indicated concentrations of HgCl$_2$ or pCMBS (Zone 1 only) for 24 hours; label was present in the medium for the last hour of the experiment. Data given as mean per cent of paired controls (n=4), variability as SE bars.

MERCURIC CHLORIDE
ZONE 1

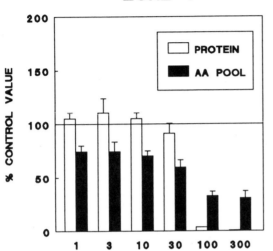

MERCURIC CHLORIDE
ZONE 2

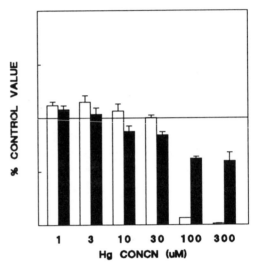

MERCURIC CHLORIDE
ZONE 3

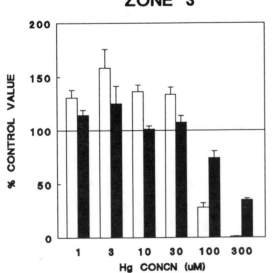

pCMBS
ZONE 1

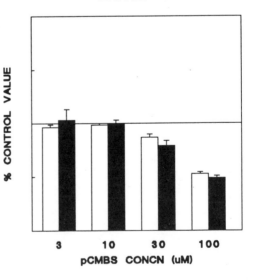

MECHANISTIC STUDIES ON THE INFLUENCE OF ADRIAMYCIN ON AMINO ACID UPTAKE AND AMINO ACID INCORPORATION INTO ISOLATED GLOMERULI OF THE ATLANTIC HAGFISH MYXINE GLUTINOSA

Sabine Kastner, Lüder M. Fels, Lena Emunds, Hilmar Stolte
Laboratory of Experimental Nephrology, Division of Nephrology, Department of Internal Medicine, Medical School Hannover, FRG

Previous metabolic studies on isolated and in vitro incubated glomeruli of the Atlantic hagfish Myxine glutinosa showed that Adriamycin (ADR) increases the incorporation of amino acids into glomerular proteins. This might be due to alterations in different pathways:
- 1. uptake of amino acids into the glomerulus
- 2. RNA-synthesis
- 3. proteolytic degradation

It is generally accepted that the cytotoxicity of the anticancer drug Adriamycin is caused by enzymatic redox cycling leading to DNA-strand breaks, peroxidation of membrane lipids and enzyme inactivation (Scheulen, M.E. et al., Archs. Toxicol. 60: 154, 1987). Therefore we also focused on the involvement of free radicals. Other mechanisms as the intercalation into DNA and RNA with subsequent inhibition of replication, transcription and translation are also discussed (Gianni, L. et al., Rev. Biochem. Toxicol. 5: 1, 1983).

Myxine was treated with 20 mg/kg b.w. ADR by injection into the caudal blood sinus. A second group of animals was treated with 20 mg/kg b.w. ADR + 450 mg/kg b.w. of the sulfhydryl-donor N-Acetylcysteine (NAC); a third group was treated with NAC alone. 10 days after treatment the glomeruli were isolated by microdissection and incubated with ^3H amino acids as described previously (Kastner, S. et al., Bull. MDIBL 29: 127, 1990). After incubation the total uptake of ^3H amino acids into the glomerulus and the incorporation of ^3H amino acids into glomerular TCA-precipitable proteins were determined. Glomerular de novo RNA-synthesis was quantified by the incorporation of 6-^3H uridine into RNA of glomeruli from controls and ADR-treated animals (Kastner, S. et al., Bull. MDIBL 30: 120, 1991).

The results showed that the amino acid uptake into glomeruli of ADR-treated animals is significantly inhibited compared to controls. Studies on glomeruli isolated from animals which received combined treatment of ADR and the radical scavenger N-Acetylcysteine revealed that NAC could prevent the inhibition of amino acid uptake after ADR-treatment (Fig. 1). 12 hours after incubation there is no significant difference in amino acid uptake between the control group (13942 ± 685 DPM/glomerulus, n=10) the NAC group (14190 ± 2074, n=10) and the ADR+NAC group (16045 ± 806, n=11). The amino acid uptake into glomeruli of ADR-treated animals is decreased to 7649 ± 1010 DPM/glomerulus (n=10). - In contrast the stimulation of amino acid incorporation into glomerular TCA-precipitable proteins could not be prevented by the radical scavenger N-Acetylcysteine. 12 hours after incubation the amino acid incorporation is not significantly different between the ADR group (7353 ± 926 DPM/glomerulus, n=27) and the ADR+NAC group (6633 ± 866, n=20) (Fig. 2). - RNA-synthesis quantified by the incorporation of 6-^3H uridine into glomerular RNA is significantly decreased after 8 hours incubation in glomeruli of ADR-treated animals (206 ± 22 DPM/glomerulus, n=26) compared to controls (438 ± 60, n= 45) (Fig. 3).

As already suggested these studies indicate that Adriamycin acts via different pathomechanisms: - 1. ADR reduced the amino acid uptake into glomerular cells. This effect is preventable by the radical scavenger N-Acetylcysteine. Therefore oxidative stress on membrane components seems to be a reasonable explanation for the inhibited amino acid uptake. Amino acid incorporation into glomerular proteins is increased. This increase is not preventable by the radical scavenger NAC. This effect is best explained by metabolic disturbances in protein synthesis and/or protein degradation. - 2. The RNA-studies showed after 8 hours incubation a significant inhibition of RNA-synthesis in glomeruli isolated from ADR-treated animals. The enhanced amino acid incorporation of glomeruli from ADR-treated animals could therefore not be explained by an enhanced RNA-synthesis. Hypothetically, increased protein synthesis could as well be due to ADR-effects on RNase activity. This could lead to a longer stability of RNAs, thus causing an increased expression of proteins. - 3. Another explanation for the accumulation of ^{3}H amino acids in glomerular proteins might be a decreased degradation caused by an inhibition of proteolytic enzymes. It has been frequently reported that ADR interferes with proteolytic activities either directly or mediated by reactive oxygen species. Therefore it remains to be further elucidated if the accumulation of ^{3}H amino acids in glomerular proteins is due to an enhanced protein synthesis or a decreased proteolytic degradation.

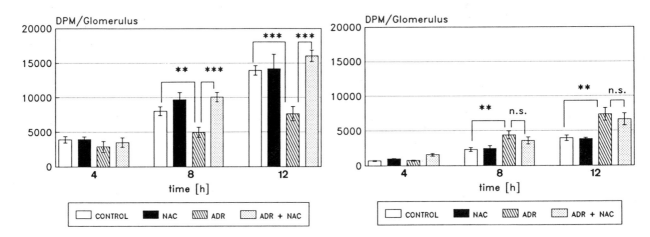

Fig. 1: Amino acid **uptake** into the glomerulus, results expressed as DPM/glomerulus ± S.E.M., statistical significance calculated by Students t-test
* p < 0.05, ** p < 0.01, *** p < 0.001, n.s. = not significant

Fig. 2: Amino acid **incorporation** into glomerular TCA-precipitable proteins (see Fig. 1)

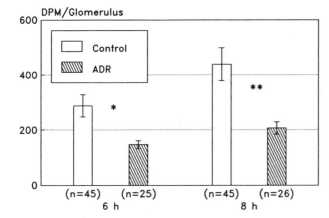

Fig. 3: Glomerular RNA synthesis, results expressed as DPM/glomerulus ± S.E.M. (see Fig. 1)

Supported by CEC (BIOT-CT91-0266) and NIEHS (EHS IP 30 ES 03828-06) to Center of Membrane Toxicity Studies

ISOLATION OF APOLIPOPROTEIN B FROM THE PLASMA OF THE DOGFISH, SQUALUS ACANTHIAS

Lorelei E. Perez, David Shultz* and Ian P. Callard
Department of Biology, Boston University
Boston, MA 02215

Lipid transport and metabolism is a dominant feature of elasmobranch physiology yet few studies have focused on this aspect of elasmobranch function. In mammalian studies, Brown and Goldstein (Science 232:34, 1986) established that the main protein component of very low density and low density lipoprotein (VLDL, LDL) fractions is apoprotein B-100, a 450 kD protein. This protein functions in the transport of the hydrophobic lipid moieties and in ligand-receptor interactions essential for lipid entry into cells. Non-mammalian species which are heavily dependant upon lipid metabolism and in which the process of yolk precursor synthesis is a major hepatic metabolic pathway, such as elasmobranchs, provide excellent models for the study of hormone regulated lipoprotein metabolism. In an earlier study Mills et al (Biochem J. 163:455, 1977) conducted a careful examination of the lipids of a shark (Centrophorus squamosus) and tentatively identified a large protein moiety of VLDL and LDL as the homologue of mammalian apolipoprotein B. As part of a research program aimed at understanding the hormonal regulation of lipoprotein synthesis and function in non-mammalian species, we sought to verify and extend the observations of Mills et al (ibid, 1977) using a readily available species, Squalus acanthias.

Dogfish were caught off the coast of Maine and maintained in flow-through circulating sea-water tanks at ambient water temperatures during July and August, 1991. Blood (10 ml) was obtained by caudal puncture and centrifuged in the presence of the protease inhibitor PMSF. The plasma was centrifuged in salt solutions sufficient to obtain VLDL (density less than 1.007 g/ml), LDL (density 1.007 - 1.065 g/ml) and HDL (density 1.065-1.21 g/ml) in an ultracentrifuge as described by Mills et al (ibid, 1977). Fractions were verified by electron microscopy. Aliquots of the isolated fractions were electrophoresed (6.0% SDS-PAGE) and transferred to nitrocellulose filters by semi-dry blotting techniques in methanol-Tris:SDS buffer. The blots were probed with a 1:1000 dilution of anti-chicken apoprotein B antibody (courtesy of Dr. D. Williams). Two sharp bands appeared on the blots, the first migrating with an approximate molecular weight of 350 kD, the second of about 50 kD. The high molecular weight protein (350 kD) corresponds well to what has been shown previously for avians (360 kD; Gehrke, L., Bast, R.E., and Iland, J. (1981) JBC 256:2522). However, the identification of the lower molecular weight component (50 kD) is not known at the present time. It is possible that this protein could be a homolog to apo B48, a truncated hepatic/intestinal version of apolipoprotein B100, or a breakdown product. It is interesting that this protein does not appear in the HDL ultracentrifugal fraction.

The results verify the work of Mills et al (ibid, 1977) and further identify the major protein component of elasmobranch VLDL and LDL as apoprotein B. It appears that common mechanisms for transport and presumably cellular uptake of lipids exist from mammals to elasmobranchs; further, the apoprotein B molecule appears to be quite conserved on the basis of the cross-reactivity of the elasmobranch protein with an antibody to chicken apoprotein B. These studies will facilitate future work on the regulation of this protein by hormones and its role in the uptake of lipids into specific cells, particularly the oocyte.

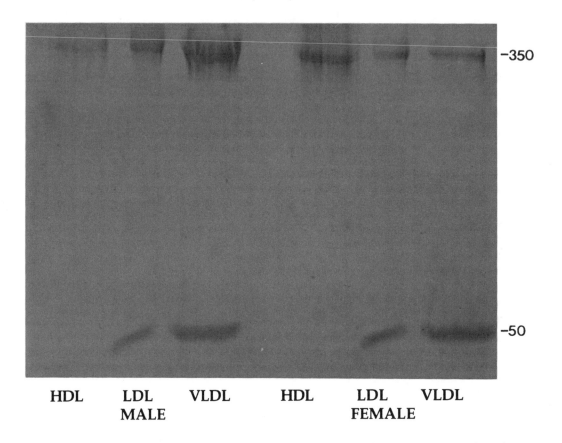

HDL	LDL	VLDL	HDL	LDL	VLDL
	MALE			FEMALE	

Figure 1 - Western blot of male and female dogfish (Squalus acanthias) isolated lipoproteins. Male and female dogfish plasma was ultracentrifuged to obtain several lipoprotein density classes. The lipoproteins were separated on 6.0 % SDS-PAGE, transferred to nitrocellulose membranes and probed with rabbit anti-chicken apolipoprotein B100 antibody. Two proteins, approximately 350 kD and 50 kD respectively, were isolated. VLDL = very-low density lipoprotein; LDL = low density lipoprotein; HDL = high density lipoprotein.

*Hearst Foundation Fellow.
Supported by NIH 1 RO1 RR 06633-01 to IPC and an NSF pre-doctoral fellowship to LEP.

38.

IDENTIFICATION OF PUTATIVE SHARK (SQUALUS ACANTHIAS) VITELLOGENIN

Lorelei E. Perez and Ian P. Callard.
Department of Biology, Boston University, 5 Cummington Street, Boston, MA 02215

Vitellogenin, a complex lipophosphoglycoprotein, is the main source of nutrients for embryos of non-mammalian, vertebrate species. The protein is synthesized in the liver, transported in the blood stream to the ovary where it is taken up via receptor-mediated endocytosis by the oocyte. Once in the ovary, it is cleaved into smaller yolk components and stored for embryonic use.

Vitellogenin has been isolated in many species, most notably Xenopus and chicken [Wallace, R.A. In: Dev Biol vol I, L. Browder, ed., Plenum, NY,1985]. However, little work has been done on the earliest gnathostomes, the elasmobranchs [Woodhead, P.M.J.,Gen Comp Endocrinol, 13:310,1969; Craik, J.C.A., Gen Comp Endocrinol, 35:455,1978]. The elasmobranchs are unique, being the earliest group of extant vertebrates to have evolved viviparity and the associated changes in vitellogenin production and development of placentae and placental analogs. We have recently described the isolation and purification of vitellogenin in the oviparous skate Raja erinacea [Perez,L.E. and Callard, I.P., in preparation, 1992a]. Here, we describe the identification of putative vitellogenin in the viviparous dogfish, Squalus acanthias. To our knowledge, this is the first report on the identification of vitellogenin in a viviparous elasmobranch.

A pregnant female shark (stage C) was bled from the caudal vein into heparinized tubes containing 1 mM PMSF (phenyl-methyl sulfonyl fluoride) and 1 mM leupeptin to inhibit protease activity. Plasma was collected after centrifugation and vitellogenin precipitated with the addition of 20 mM EDTA and 0.5 M $MgCl_2$. All procedures were performed at 4 °C. The mixture was incubated overnight, and centrifuged for 15 minutes at 2500 x g. The pellet was resuspended in 3 ml 1 M NaCl/50 mM Tris-HCl, pH=7.5 and centrifuged to remove debris. The supernatant was re-precipitated with the addition of 8 volumes of water containing 1 mM PMSF/1 mM leupeptin. The mixture was incubated overnight and the precipitate centrifuged at 2500 x g for 15 minutes. The pellet was resuspended in 3 ml 1M NaCl/50 mM Tris-HCl, pH=7.5, de-salted and concentrated by centrifugation on Centricon-100 (Amicon) filtration tubes, lyophillized and stored at -70 °C. Samples were run on 0.75 mm, 6% SDS-PAGE using the buffer system of Laemmli [Nature, 227:680, 1970]. Gels were stained in 0.2% Coomassie blue for visualization of proteins. Duplicate gels were transferred onto nitrocellulose membranes and probed with rabbit anti-skate vitellogenin antibody [Perez L.E. and Callard, I.P., in preparation, 1992b].

The addition of MgCl$_2$/EDTA precipitated a wide spectrum of proteins as seen on polyacrylamide gels. However, 3 prominent high molecular weight bands at 241, 225, and 207 kD respectively appeared in female plasma. No proteins were precipitated by MgCl2/EDTA from male plasma (data not shown). Western blot analysis of female dogfish precipitates with anti-skate vitellogenin showed weak, binding with only the high molecular weight triplet proteins (data not shown). Interestingly, female skate plasma probed with rabbit anti-skate vitellogenin antibody exhibits strong specific binding with only one protein band at 205 kD [Perez L.E. and Callard, I.P., in preparation, 1992b]. Problems with degradation of precipitated proteins occurred thoughout the procedure, even in the presence of PMSF and leupeptin, which is common when working with this protein (see Wallace, 1985, ibid). This may explain the presence of smaller molecular weight proteins near the dye front in the gel, although we cannot rule out the possibility of non-selective precipitation. Further purification steps utilizing DEAE-cellulose chromatography resulted in complete degradation of the triplet protein.

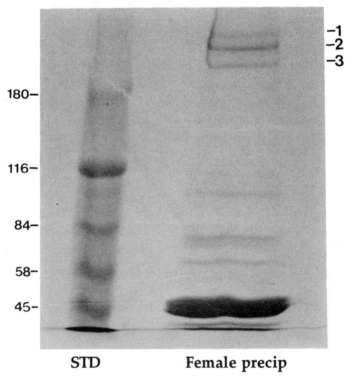

FIGURE 1 - MgCl2/EDTA precipitation of female dogfish (Squalus acanthias) plasma. A triplet of high molecular weight proteins (241, 225, 207) kD, which weakly binds to anti-skate vitellogenin antibody, is precipitated with this technique. STD = standard; Female precip = female dogfish MgCl2/EDTA precipitate.

Taken together, the data suggest that the triplet protein isolated from female dogfish plasma has similar biochemical characteristics to vitellogenin, being selectively precipitated with $MgCl_2$/EDTA. This technique has previously been used to precipitate <u>Xenopus</u>, <u>Rana</u>, <u>Gallus</u>, and a variety of other vertebrate vitellogenins [Wiley <u>et al</u>, Anal. Biochem., 97:48,1980; Carnevali and Polzonetti-Magni, J. Exper. Zool., 259:18,1991]. In addition, anti-skate vitellogenin antibody weakly, but specifically, recognizes the triplet protein in Western blots. In the appearance of a triplet of proteins, the dogfish differs from the oviparous <u>Raja erinacea</u>, where only one vitellogenin subunit could be detected. This suggests that vitellogenin has structurally evolved within the elasmobranch group. These differences may relate to the evolution of viviparity and associated changes in vitellogenin genes.

Supported by NIH 1 RO1 RR 06633-01 to IPC and an NSF pre-doctoral fellowship to LEP.

EFFECTS OF HEAVY METALS ON DNA SYNTHESIS IN THE TESTIS OF THE DOGFISH (SQUALUS ACANTHIAS)

J. Michael Redding
Department of Biology, Tennessee Tech University, Cookeville, TN 38501

Contamination of aquatic environments with metal compounds is a serious risk to the health of aquatic species and terrestial species that rely on food from aquatic environments. Widespread metal contamination of both marine and freshwater systems has been reported, and there is a large body of literature documenting deleterious effects of such pollution on various species. Of particular concern is the tendency of animals, especially carnivores, to accumulate metals from dietary sources, thereby increasing their risk for dose dependent toxic effects. The spiny dogfish, Squalus acanthias, and other elasmobranchs are especially susceptible to such cumulative effects because they are long-lived, carnivorous animals whose home range includes coastal marine habitats which tend to have the highest concentrations of metals. Several studies have documented high metal concentrations in tissues of the spiny dogfish and other members of the genus Squalus (e.g., Taguchi, Mar. Environ. Res. 2: 239, 1979). Metal intoxication may be directly detrimental to the health of the dogfish populations by decreasing survival or reproductive fitness.

The known toxic effects of metals are diverse. Effects are thought to be exerted via the formation of stable complexes with many different biological molecules including proteins, DNA, RNA, and phosphorylated compounds. Specific mechanisms of metal toxicity have been characterized extensively for mammalian systems, but much less is known about non-mammalian systems. Information on the specific cellular effects of metals on male reproductive systems in vertebrates is sparse but suggestive of profound disturbances (Clarkson, et al., Reproductive and Developmental Toxicity of Metals, Plenum Press, 1983; Mottet and Landolt, Environ. Health Perspectives 71: 69, 1987).

In recent years, the dogfish testis has proven to be an excellent model for studying the regulation of vertebrate spermatogenesis (Callard et al., J. Exp. Zool. Suppl. 2: 23-34, 1989). Distinct developmental stages of spermatocysts (germ cell:Sertoli cell units) can be isolated and cultured in vitro for at least two weeks (Callard and Dubois, Bull MDIBL 27: 30, 1988). Moreover, mitotic activity, as indicated by DNA synthesis, is maintained quantitatively during this period and is responsive to stimulatory and inhibitory factors (Redding and Callard, Bull. MDIBL 30: 30-32, 1991). Thus, this model system would seem suitable for toxicological studies of vertebrate spermatogenesis. The purpose of this study was to determine the effects of various metals on DNA synthesis in the testis of the spiny dogfish.

For each experiment, spermatocysts were isolated from zone I tissue from testes of 2-4 sharks and maintained in culture with Leibovitz L-15 medium, modified for use with elasmobranch tissue as noted in Redding and Callard (op. cit.). After various periods of treatment with metals, 5.0 uCi/ml of ^{3}H-thymidine was added to the cyst cultures for 6-24 hr before harvesting the cysts. Harvested cysts were washed twice with saline solution augmented with excess unlabelled thymidine. Then, cysts were treated with ice-cold 10% trichloroacetic acid for 1-24 hr. Cysts were then washed again before solubilizing overnight in 0.2 M NaOH. Aliquots of the solubilized cysts were analyzed for radioactivity (cpm). Data were not standardized by sample protein concentration as previously reported; in these experiments such standardization would not significantly change the results. Results from cysts treated with metals were standardized as a percentage of the mean of untreated controls. The standardized means of treatment groups were compared to that of untreated control groups by an unpaired t-test with a pooled variance estimate.

Preliminary experiments showed that mercuric chloride ($HgCl_2$) at concentrations greater than 100 uM inhibited synthesis of DNA. Subsequently, the effects of equivalent concentrations (100, 500, 1000 uM) of $HgCl_2$, the organic mercurial parachloro-mercuric-phenol-sulfonic acid (PCMBS), sodium vanadate ($NaVO_3$), zinc chloride ($ZnCl_2$), and cadmium chloride ($CdCl_2$) were evaluated. Effects of these metals were compared to untreated controls, positive controls treated with bovine insulin (10 ug/ml), and negative controls treated with isobutyl-methylxanthine (1 mM, IBMX). Results of this experiment are shown in table 1.

Table 1. DNA synthesis rates of <u>Squalus</u> zone I spermatocysts cultured in vitro with metals, insulin (10 ug/ml, positive control), or IBMX (1 mM, negative control) for 24 hr and exposed for the last 12 hr to radiolabelled thymidine. Results are shown as a mean (SE) percentage of control cultures. Sample size was four for each treatment and eight for the untreated control. All means except those noted by "ns" were significantly (P< 0.01) different from the control.

| | | Metal Concentration (uM) | | |
Treatment	0	100	500	1000
Control	100 (2)	---	---	---
Insulin	192 (7)	---	---	---
IBMX	34 (3)	---	---	---
$HgCl_2$	---	84 (2)	3 (1)	0 (0)
PCMBS	---	95 (3)ns	62 (3)	8 (3)
$CdCl_2$	---	128 (6)	69 (2)	21 (3)
$NaVO_3$	---	98 (3)ns	---	71 (4)
$ZnCl_2$	---	117 (6)	117 (6)	127 (12)

Mercurial compounds showed dose dependent inhibition of DNA synthesis. Of these $HgCl_2$ was most potent, virtually negating DNA synthesis between 100 and 500 uM. Cadmium stimulated DNA synthesis at 100 uM but markedly inhibited it at 500 and 1000 uM. Vanadate was relatively ineffective, reducing DNA synthesis only to 71% of controls at 1000 uM. In contrast, zinc slightly stimulated DNA synthesis at all concentrations. Beyond simply identifying their effective concentrations, these results demonstrate the specificity of various metals with respect to their effects on DNA synthesis in shark testis. This specificity may reflect differences in the capacity and affinity of metal binding proteins such as metallothionine that effectively sequester metals and prevent them from affecting critical cellular processes. It is evident from these results that DNA synthesis in the testis is sensitive to metal intoxication, generally supporting previous results from mammalian models (see Clarkson et al., op. cit.). These results support the use of <u>Squalus</u> testis as a model system for toxicological studies of vertebrate spermatogenesis.

[This research was supported by a fellowship from the Lucille P. Markey Charitable Trust. I thank G.V. Callard, D. S. Miller, D. Barnes and A. Kleinzeller for valuable advice and material support.]

NITROGEN BUDGET IN DEVELOPING ELASMOBRANCH EMBRYOS

Gregg A. Kormanik[1], Antoine Lofton[1] and Niamh O'Leary-Liu[2]
[1]Department of Biology, Univ. of North Carolina at Asheville, NC 28804
[2]P.O. Box 218, Surrey, ME 04684.

We have shown that late-term embryos of the dogfish Squalus acanthias are retained in a uterine environment relatively high in ammonia, and tentatively suggested that this ammonia may act as a nitrogen source for the developing embryos (Kormanik & Evans, J. Exp. Biol. 125;173-179, 1986). A subsequent investigation showed that total nitrogen declines by about 16% when stage 'A' embryos are compared to stage 'C' embryos (Kormanik, J. Exp. Biol. 144;583-587, 1989). While this decline is significant, the intriguing aspect is that the decline in nitrogen is not as great as the total decline in dry weight (ca. 40%, Hisaw & Albert, Biol. Bull. 92;187-199, 1947). This rather modest loss of nitrogen may in fact represent some nitrogen transfer from mother to embryo. To extend these previous observations, we investigated the nitrogen budget in an oviparous species, Raja erinacea, where there is no maternal contribution.

Dogfish (Squalus acanthias) and skate (Raja erinacea) embryos were collected as previously described (Kormanik & Evans, ibid.; Kormanik et al., this issue). Eggs were opened, the egg cell (= yolk) and jelly were separated, preserved with sulfuric acid (final conc. 10%) and frozen. Remaining eggs were labelled, kept in sea water and returned to University of North Carolina at Asheville where they were incubated at about 25° C. until hatching (ca. 7 mo.). Hatchlings were killed by decapitation and homogenized. Total nitrogen (N_{tot}) was determined in embryos and egg contents using a Hach Digesdahl, according to standard procedures (Hach).

In another series of experiments we determined the egg and embryo content of urea and trimethylamine oxide (TMAO), the major nitrogenous osmolytes in the elasmobranchs. Fresh eggs were opened, portions of the yolks were homogenized on ice with distilled H_2O (1:9), precipitated with trichloroacetic acid (final conc. 3.5%) and centrifuged. Embryos were killed by decapitation, homogenized on ice and otherwise treated as the egg homogenates. Samples were analyzed for urea (Sigma Kit # 535). TMAO was assayed using a procedure modified from Forster et al. (J. Gen. Physiol. 42:2;319-327, 1958). Samples of the yolk were dried to constant weight at 60° C. to determine water content. Results are expressed as mean ± 1 S.E.M.

The results of the N_{tot} determinations are presented in Table 1. In preliminary experiments, we found that about 98% of the N_{tot} was in the egg cell; little was found in the jelly. The sum of both components is included in Table 1. N_{tot} decreased by nearly 40%. The loss of nitrogen in this oviparous species is more than twice as great as that observed for S. acanthias. N_{tot} per unit wet weight decreased as the embryos increased in weight.

The results of the urea and TMAO determinations are presented in Table 2. Urea and TMAO concentrations in the egg cells of the skate are higher than those of adult plasma (340 and 50 mM, respectively; this investigation). The distribution and concentrations are similar to those seen in muscle cells of this species (Forster & Goldstein, Am. J. Physiol.

Table 1. Nitrogen in eggs and hatchlings of the skate, Raja erinacea.

	wet wt. (g)	N_{tot} (mg)	N_{tot}/g (mg/g wet wt)
Egg (n = 10-11)	4.32 ± 0.29	283 ± 17	66.0 ± 1.5
Hatchlings (n = 5)	5.97 ± 0.49	171 ± 14	26.9 ± 4.6
Difference (Hatchlings/eggs)	138%	60.4%	40.8%
Signif. (p <) (2-tail)	0.01	0.001	0.001

Table 2. Urea and TMAO in egg cells of the skate, Raja erinacea (n = 5-6).

	Urea	TMAO	H_2O (in %)
Amount [$umol*g^{-1}$]	230 ± 10	96.4 ± 2.7	57.8 ± 0.7
Concentration [$mmol*(kg\ H_2O^{-1})$]	398 ± 21	167. ± 5	
% egg N_{tot}	9.8	2.0	

230(4):925-931, 1976). The amount of nitrogen as urea and TMAO in the eggs represents less than 12% of the total nitrogen stores.

In both species, losses of urea, ammonia and TMAO via the gills and kidney probably represent the major routes of nitrogen loss.   Rates of ammonia and urea excretion are similar in these two species (Evans & Kormanik, J. Exp. Biol. 119:375-380, 1985; Payan et al., Am. J. Physiol. 224;2,367-372, 1973).   Certainly urea (Mommsen and Walsh, Science 243;72-75, 1989) and possibly TMAO (Read, Biol. Bull. 135:3;537-547, 1968) are synthesized to replace that lost by excretion and diffusion. However, Goldstein et al. (Comp. Biochem. Physiol. 21;719-722, 1967) were unable to demonstrate TMAO synthesis in S. acanthias.   We examined the changes that occur in these pools in S. acanthias, since R. erinacea hatchlings were unavailable for this series of experiments.   The results are presented in Table 3.

Urea and TMAO contents of early-term S. acanthias embryos are similar to those of R. erinacea.   Since eggs and embryos vary in size (see below), and we are unable to track specific eggs as they develop, the best we can do is to compare the ranges, and assume the smallest or largest eggs give rise to the smallest and largest of embryos. As the embryos develop, then, the wet weight increases, total TMAO decreases, and total urea increases. Total nitrogen, treated in the same manner, decreases only 3 to 14 %.   Since the uterine environment for these late-term dogfish embryos contains little urea

Table 3. Urea and TMAO content in early and late-term embryos of the dogfish, Squalus acanthias. Early-term and late-term are stages 'A' and 'C' of Hisaw & Albert (ibid.), about one year apart in development.

Content	wet wt. (g)	TMAO	Urea $(umol\ g^{-1})$	N_{tot}[1]
early-term	30.6 ± 1.8 (18)	110 ± 8 (3)	202 ± 3 (17)	3450
late-term	39.2 ± 2.8 (6)	67.8 ± 1.4 (6)	272 ± 19 (5)	2400

Total Content	wt. range[2] (g)	TMAO range	Urea range (mmol)	N_{tot} range
early-term	21 - 43	2.3 - 4.7	4.2 - 8.6	72 - 148
late-term	26 - 60	1.8 - 4.1	7.1 - 16.3	62 - 144
(late-/ early-term)	(124 - 140%)	(78 - 87%)	(169 - 189%)	(86 - 97%)

1 - N_{tot} from Kormanik, J. Exp. Biol. 144;583-587, 1989.
2 - from Kormanik et al., unpubl. and Hisaw and Albert (ibid.)

or TMAO, these data demonstrate that urea is synthesized from nitrogen stores (endogenous or exogenous) as the embryo develops, but TMAO, stored in the egg, declines and is not replaced by synthesis (Goldstein et al., ibid.).

These data show that embryos of the skate, R. erinacea, an oviparous species, lose more nitrogen during development than do embryos of the primitively viviparous dogfish, S. acanthias. This piece of evidence, albeit indirect, supports a role for uterine ammonia in the nitrogen budget of S. acanthias. The dogfish embryo must synthesize urea, since the urea pool increases through development, as the pool of total nitrogen decreases. Dogfish embryos lose 23 mmol N in about 1 year (Kormanik, ibid.), while the urea pool increases by 4.5 mmol (from Table 3). If all of this N is considered as urea (16 mmol), embryos need to synthesize at most about 2 umol urea embryo^{-1} hour^{-1} to account for the urea accumulated and/or lost by diffusion, a number well within the synthetic capabilities of elasmobranch tissues (Anderson, Science 208;291-293, 1980; Mommsen and Walsh, ibid.). Glutamine is the donor of $-NH_2$ groups to CPS III of the ornithine-urea cycle, which may also accept ammonia albeit at a lower rate (Anderson, ibid.; Mommsen and Walsh, ibid.). Blood ammonia in late-term uterine incubated embryos is 3-4 fold higher than fish in fresh sea water (Kormanik, J. Exp. Biol. 137;453-456, 1988). The potential effect of this elevated ammonia on urea or glutamine synthesis should be considered. (Supported by NSF DCB-8904429 to GAK and a Hearst Foundation, Inc. scholarship to NOL).

STAGE-DEPENDENCY OF PROTEINS AND GELATINOLYTIC PROTEINASE ACTIVITIES DURING SPERMATOGENESIS

Marlies Betka [1], Joseph E. Haverly [2] and Gloria V. Callard [1]
[1] Department of Biology, Boston University, Boston MA 02215
[2] Department of Zoology, Washington State University, Pullman WA 99163

Spermatogenesis is an unique developmental process involving profound changes in number and shape of germ cells and associated somatic elements (Sertoli cells). Presumably, it requires the turning on and off of different genes in a strict temporal and cell-specific order, which should be reflected in changes in the protein products of gene expression. Due to the complex organization of the mammalian testis, it is not feasible technically to analyze gene expression or protein synthesis stage-by-stage in intact germinal units. In the testis of the spiny dogfish (Squalus acanthias), however, the primary germinal units (spermatocysts: isogenetic germ cell clones plus stage-synchronized Sertoli cells) are anatomically distinct and can be isolated intact (Callard et al., J. Exp. Zool. Suppl.2: 23, 1989). Moreover, spermatocysts are arranged in maturational order across the diameter of the testis, facilitating analysis of proteins in relation to specific developmental stages. We report here initial studies in which we document protein composition of premeiotic (PrM), meiotic (M) and postmeiotic (PoM) spermatocysts.

Testes of 3-5 dogfish were dissected according to stage of development and spermatocysts isolated according to procedures described by Dubois & Callard (J. Exp. Zool. 258: 359, 1991). To solubilize proteins, cysts (20 μl packed volume) were boiled for 30 min in 30 μl 20% SDS and an additional 10 min after adding 270 μl buffer (0.065 M Tris-HCl, 10% glycerol, 2% ß-mercaptoethanol, 0.04% bromophenol blue). After centrifugation for 15 min at 12,000 x g, protein concentrations were determined in TCA-precipitated aliquots by the method of Lowry et al. (J. Biol. Chem. 193: 265, 1951). Proteins (100 μg/well) were separated by SDS-PAGE (Laemmli, Nature 277: 680, 1970) using 4-20% linear gradient gels with a 4% stacking gel, followed by Coomassie blue staining. Apparent molecular weights (Mr) of stained bands were estimated by linear regression analysis using known molecular weight markers as standards. The relative abundance of a given protein band was estimated visually by staining intensity using a scale of 0 to +++, with + being assigned to the stage of lowest intensity and 0 indicating no detectable staining. In addition, comparisons were made of all proteins in a given stage, and those of greatest intensity indicated.

The stage-related distribution of 13 representative proteins of a total of 30-35 observed bands is summarized in Table 1 for spermatocysts isolated in July versus September, and results were similar for at least 4 other preparations. Most protein bands were present in all stages but each displayed a unique stage-dependent pattern. For example, in July the 28.6 kDa band was PrM > M > PoM, whereas the 109.8 kDa band was PoM > M = PrM and the 54.8 kDa band was M > PrM > PoM. Only a few bands were stage-specific (e.g. 90.8 kDa in PoM stages only) or were equivalent in all stages (e.g. 83.9 kDa band). Comparing spermatocysts from July versus September, which represent the progression toward spermatogenic inactivity, some changes were detectable (e.g. the 35.7 kDa band appears only in September), but the changes were minor for most protein bands.

Table 1. SDS-PAGE analysis of proteins in premeiotic (PrM), meiotic (M) and postmeiotic (PoM) spermatocysts
The grading (0,+,++,+++) reflects the intensity of a given protein across all three developmental stages. The most abundant proteins within each stage are marked with *.

Apparent Mr kDa	July			September		
	PrM	M	PoM	PrM	M	PoM
109.8	+	+	++	+	+	++
98.3	++	++	+	++	++	+
90.8	0	0	+	0	0	+
72.8	+	++	+	0	++	+
83.9	+	+	+	0	+	+
54.8	++*	+++*	+*	++*	++*	+*
44.6	+	+	+	+	+	+
35.7	0	0	0	+	+	0
28.6	+++	++	+	+++*	++*	+
25.2	0	+	+	+	++	+++*
16.5	++*	++*	+	++	++*	+
15.7	++*	++*	+	++*	++*	+
14.5	++*	++*	+	++	++*	+

Because the profound remodeling of both germ cells and Sertoli cells which occurs during spermatogenesis might involve limited proteolysis, we used the technique of gelatin-containing SDS-PAGE zymography (Heussen & Dowdle, Anal. Biochem. 102: 186, 1980) to determine whether any of the observed protein bands were associated with proteinase activity. In this procedure, proteinase activity is indicated on electrophoretograms as clear bands on a uniformly amido black-stained background of gelatin. We found proteinases with apparent Mr of 104 kDa, 81 kDa and 69 kDa in all three spermatogenic stages and each was uniquely stage-related. The 104 kDa proteinase showed the same intensity throughout all three stages, while the activity of the 81 kDa proteinase was highest in PrM and PoM cysts. The enzymatic activity of the 69 kDa proteinase was clearly visible in PoM cysts but very low in the other stages; however, this distribution of activity varied between different cyst preparations. It remains to be determined whether these proteinase activities correspond to Coomassie blue stained bands with Mr of 109.8 kDa, 83.9 kDa and 72.8 kDa.

Initial results demonstrate the feasibility of this approach for analysis of changes in cyst protein composition during spermatogenesis and as a first step in identifying specific proteins and their genes. Furthermore, because staged spermatocysts are amenable to culture for up to 7 days (DuBois and Callard, J. Exp. Zool. 258: 359,1991), future experiments will allow radiolabeling of synthesized proteins directly and the testing of putative regulators of gene expression (e.g. steroids and other hormones, growth factors) under defined conditions in vitro.

(Supported by a grant from NIH HD16715 to GVC and a fellowship from the Pew Charitable Trust to JEH).

PARACRINE REGULATION OF PREMEIOTIC GERM CELL PROLIFERATION IN THE TESTIS OF THE SPINY DOGFISH, SQUALUS ACANTHIAS

Francesc Piferrer and Gloria V. Callard
Department of Biology, Boston University, Boston, MA 02215

The regulation of spermatogenesis is a complex process involving functional interactions among several cell types. Studies trying to elucidate the underlying mechanisms are made difficult by the complex organization of the mammalian testis. In contrast, the dogfish shark, Squalus acanthias, has testes in which germ cells plus associated somatic cells form discrete anatomical units (spermatocysts) and are arranged in maturational stages across the diameter of the testis. Using the shark testis model (see Callard et al., 1989, J. Exp. Zool., Suppl. 2: 353-364, for review), it was shown that a basic cell function, DNA synthesis, can be maintained quantitatively for several days under in vitro conditions and is stage-dependent. DNA synthesis by premeiotic (PrM) spermatocysts is 6-fold higher than that observed in meiotic (M) and postmeiotic (PoM) spermatocysts (DuBois and Callard, 1991, J. Exp. Zool., 258: 359-372). As part of studies in which we are surveying possible regulators of DNA synthesis in PrM cysts, we tested the hypothesis that more mature developmental stages, which are upstream in the vascular pathway within the testis, may signal the developmental advance of less mature stages.

PrM spermatocysts were cultured in the inner well of a two-chamber microwell assembly for 48h (t_0 to t_{48}) in the presence of medium alone or with PrM, M or PoM spermatocysts in the outer well. DNA synthesis was measured from t_{24} to t_{48} by the incorporation of [methyl-3H] thymidine into trichloroacetic acid (TCA) precipitable material. Radioactivity per unit cyst protein (micro-Lowry) was not significantly altered when PrM spermatocysts were cocultured with either PrM or M spermatocysts but decreased by one-half in the presence of PoM spermatocysts (Table 1A). The inhibition of DNA synthesis in PrM cysts caused by PoM cysts was confirmed in a dose-response study. Coculture of PrM cysts with 1 mg excess cysts of the same stage (PrM) reduced DNA synthesis at most to 66% of the control values. In contrast, an excess of 0.5 mg PoM spermatocysts reduced DNA synthesis in PrM spermatocysts to 12% of controls (Table 1B).

Further investigations tested the specificity of the response by including other tissues. DNA synthesis was not affected when PrM cysts were cocultured with different amounts of muscle, epididymis or PrM fragments, whereas fragments of PoM tissue and epigonal organ (a lymphomyeloid tissue encapsulating the testis immediately adjacent to the PoM region) markedly reduced DNA synthesis in PrM cysts. These results were confirmed in a second, similar experiment and also when tissue extracts rather than tissue fragments were added to PrM cysts. Further, in contrast to the inhibitory effects of testicular or epigonal tissue, a powerful stimulatory effect on PrM DNA synthesis was observed when extracts of spermatozoa collected from the sperm sac were tested.

Table 1. Regulation of DNA synthesis in premeiotic (PrM) spermatocysts.

A) Effects of coculture with premeiotic (PrM), meiotic (M) and postmeiotic (PoM) spermatocysts (dpm/ug protein ± SEM of two separate experiments, each in duplicate)

Medium (Ctrl)	+PrM cysts	+M cysts	+PoM cysts
1260±75	1080±115	965±160	615±57

B) Dose-response effects of PrM, M or PoM spermatocysts (percentage of control in two separate experiments each in triplicate)

	Excess protein (ug)	% DNA synthesis
PrM cysts	0.0	100.0
	115.0	75.5
	282.0	67.3
	555.5	68.2
M cysts	0.0	100.0
	263.5	47.5
	470.0	32.7
	1055.5	33.6
PoM cysts	0.0	100.0
	261.0	23.3
	466.5	12.0
	2247.5	19.2

Our findings, therefore, point to the existence of one or more substances present in the PoM region of the testis and in the epigonal organ which are capable of inhibiting the synthesis of DNA in PrM stages of spermatogenesis. These results are consistent with an early report of an inhibitory chalone in rat testis (Clermont and Mauger, 1974, Cell Tissue Kinet., 7: 165-172); however, the latter has not been further studied. Also our test system has identified a stimulatory substance associated with spermatozoa after spermiation. These stimulatory and inhibitory effects support the existence of a paracrine mechanism for the regulation of mitotic proliferation of germ cells during the early stages of spermatogenesis and provide a useful assay for isolation and purification studies of putative regulators.

(Funded by a grant from NIH HD16715 to GVC. FP supported by a postdoctoral fellowship from the Spanish government)

THE RESPONSE TO METABOLIC ACIDOSIS IN THE
MARINE TELEOST, <u>MYOXOCEPHALUS OCTODECIMSPINOSUS</u>:
FAILURE TO BE ALTERED BY CARBONIC ANHYDRASE INHIBITION

Thomas H. Maren[1] and Deborah Rothman[2]
[1]Department of Pharmacology and Therapeutics
University of Florida College of Medicine, Gainesville, FL 32610
[2]Department of Pediatrics
Boston University School of Medicine, Boston, MA 02215

This is the third and last report of our work on acid-base physiology in the representative marine teleost, <u>M. octodecimspinosus,</u> the long-horn sculpin (Bull. M.D.I.B.L. 29:62, 1990; ibid 30:104, 1991; Am. J. Physiol, Renal Fluid and Electrolyte Physiology, In Press). Here we address the question of whether the recovery from an acid load, which is largely a function of the gill, is mediated by carbonic anhydrase. We have previously shown that this enzyme at the gill is involved in the rapid recovery from metabolic alkalosis, i.e. injection of 8 meq/kg $NaHCO_3$ (Bull. M.D.I.B.L. 30:104, 1991).

In previous experiments in <u>Squalus acanthias</u> Claiborne and Swenson showed that the recovery from metabolic acidosis was unaffected by carbonic anhydrase inhibition at the gill (Bull. M.D.I.B.L. 26:5, 1986). At first consideration this seemed a surprising result. In any case, we thought it important to run similar experiments in the teleost.

The methods have been described (Bull. M.D.I.B.L. 29:62, 1990) and the response to acidosis (0.7 meq/kg HCl) in terms of blood acid-base balance and urinary excretion reported (Bull. M.D.I.B.L. 30:104, 1991). We now report the effect of carbonic anhydrase inhibition at the gill brought about by intravenous injection of 1 mg/kg benzolamide (Swenson and Maren, Am. J. Physiol. 253:R450, 1987) at the same time as injection of the acid load.

Table 1, line 1 shows normal values for urine and plasma electrolytes. The pH is lower and the pCO_2 higher than generally reported (pH = 7.8 and pCO_2 = 3) for blood drawn by indwelling catheter, because of handling. Administration of acid causes a profound early drop in plasma HCO_3^- (Fig. 1A) and increase in urinary acid and phosphate (Table 1, line 2). We previously calculated (Bull. M.D.I.B.L. 30:104, 1991 and Am. J. Physiol., In Press) that 26% of the injected acid is excreted in the urine so that 3/4 must be buffered internally or escapes across the gills. The former appears unlikely, and we assume the latter, loss through the gills, along with McDonald et al. (J. Exp. Biol. 98:403, 1982) and Cameron and Kormanik (J. Exp. Biol. 99:127, 1982). We did not study the urine in the HCl + benzolamide series (Line 3 of Table 1) since there is no renal carbonic anhydrase; thus the data are assumed to be as in Line 2. Plasma values at 1 hour show metabolic acidosis, with added slight respiratory component in the presence of the carbonic anhydrase inhibitor (see below.)

Figure 1A shows the rapid (4-hour) recovery from profound metabolic acidosis is terms of plasma HCO_3^- concentration. Fig. 1B shows that this in unaffected by benzolamide (note also three additional fish received 30 mg/kg methazolamide, an inhibitor with quite different properties) (Swenson and Maren, ibid) giving the same result.

Table 2 gives the changes in terms of plasma pH; again recovery is rapid and unaffected by benzolamide; note two-hour data. The lower initial pH in the benzolamide series compared to HCl alone is due to respiratory acidosis, which is usually minimal with this compound. In August 1991, however, fish were rather inactive, so the pCO_2 response is quite reasonable. This does not affect our primary finding that pH recovery is essentially normal in the inhibited fish.

TABLE 1. ACIDOSIS IN THE SCULPIN: EFFECT OF CARBONIC ANHYDRASE INHIBITION

	Flow ml/kg·h	pH	CO$_2$	TA	PO$_4$	HCO$_3^-$ mM ± S.E.$_m$	pH	pCO$_2$ mm Hg
				0 - 4 hr URINE millimolar ± S.E.$_m$		Plasma at 1 Hour		
Control (n=15)	1.4	6.6	3 ± 1	11 ± 2	12 ± 3	5.6 ± .2	7.5	5 ± .3
HCl, 0.7 meq/kg (n=6)	1.2	6.1	< 1	39	29	3.5	7.24	6
HCl + 1 mg/kg (n=4) benzolamide	—	—	—	—	—	3.5	6.96	10

Data in Lines 2 and 3 are significantly different from Line 1 (P ~ 0.02) except for flow.

The facts, now established, that carbonic anhydrase inhibition retards recovery from metabolic alkalosis but not acidosis in marine fish, are not really surprising. HCO$_3^-$ ion in the blood is excreted as CO$_2$ across lung or gill. The <u>normal</u> conversion from ion to gas is catalyzed by carbonic anhydrase in red cells (all vertebrates) but in the fish, excretion of <u>excess</u> HCO$_3^-$ loads is mediated by carbonic anhydrase in gill (Swenson and Maren, ibid). Figure 2 (left) shows this in summary form. Figure 2 (right) shows the reformation of HCO$_3^-$ in gill epithelium and mucus boundary, which is then available for various ion movements and exchange, depending on the environment.

On the other hand, Fig. 2 yields no suggestion of how a H$^+$ excess in the blood might be handled by any phase of the carbonic anhydrase system. We conclude that H$^+$ excess is dissipated by diffusion across the gill and/or internal buffering (McDonald et al., ibid) and is qualitatively different from the recovery process for HCO$_3^-$ excess.

Figure 1.

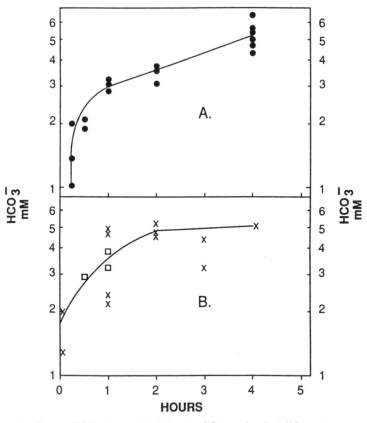

A. Plasma HCO$_3^-$ in sculpin following 0.7 meq/kg (i.v.) HCl at zero time.

B. As A but also given methazolamide, □, (30 mg/kg) or benzolamide, X, (1 mg/kg) i.v. at zero time.

TABLE 2. TIME COURSE OF PLASMA pH AND pCO_2 FOLLOWING HCl IN SCULPIN: EFFECT OF CARBONIC ANHYDRASE INHIBITION

		5 - 30 min	1 hr	2 hr	4 hr
HCl (n = 6)	pH	7.00	7.24	7.28	7.41
	pCO_2	6	6	6	6
HCl + Benzolamide	pH	6.6	6.96	7.31	7.30
(n = 7)	pCO_2	11	10	7	9

FIGURE 2. HCO_3^-/CO_2 PATHWAYS AND CARBONIC ANHYDRASE DISTRIBUTION (•••) IN FISH, RED CELL AND GILL.

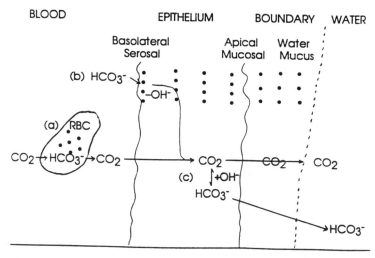

(a) Normal pathway for CO_2 elimination, blocked by methazolamide in red cells.

(b) Pathway for excess HCO_3^- load, blocked by benzolamide and methazolamide in gill.

(c) Formation of HCO_3^- in gill, to subserve various ion movements depending on environment. Adapted from Perry and Laurent, Comp. Physiol. 6:39, 1990.

Funds were provided by the Division of Sponsored Research, University of Florida, and the Alcon Research Institute.

REGULATION OF ACID-BASE BALANCE IN THE LONG-HORNED SCULPIN (MYOXOCEPHALUS OCTODECIMSPINOSUS) FOLLOWING ACID INFUSION: EFFECT OF AMBIENT SALINITY

James B. Claiborne[1] and Erin Perry[2]

[1]Department of Biology, Georgia Southern University, Statesboro, GA 30460

[2]The Mount Desert Island Biological Laboratory, Salsbury Cove ME 04672

We have previously shown that acid-base transfers in the long horned sculpin (Myoxocephalus octodecimspinosus) are impaired when the fish is exposed to dilutions of the ambient water (Walton & Claiborne, Bull MDIBL 27:4-5, 1988; Claiborne & Perry, Bull MDIBL 30:60-61, 1991). When in seawater, the sculpin is able to excrete an administered NH_4^+, HCO_3^- (Claiborne & Evans, J. Exp. Biol 140:89-105, 1988) or H^+ (Maren & Fine, Bull MDIBL 30:60-61, 1991) load mainly via the gills. Thus, while this species may possess the branchial mechanisms for acid-base regulation, (Na^+/NH_4^+, Na^+/H^+, and/or Cl^-/HCO_3^- exchange; Evans, in "Fish Physiology", eds. W. S. Hoar & D. J. Randall, Vol Xb, pp. 239-283, 1984), low external salt concentrations should alter the ability of the animal to maintain normal H^+ excretion (Claiborne & Perry, ibid.). The purpose of the present study was two-fold: (1) to measure the time course of plasma acid-base balance and net transfers between the fish and water following an acid infusion (2 meq kg^{-1} HCl), and (2) to test the effects of low salinity exposure on these parameters subsequent to the acid load.

Long-horned sculpin (Myoxocephalus octodecimspinosus) were cannulated and placed in experimental chambers according to the methods described by Walton & Claiborne (ibid.). In addition, an intraperitoneal cannula (PE-50) was introduced into the animal (Claiborne & Evans, ibid.) to allow the infusion of acid (0.1 N HCl; 2 meq kg^{-1} in teleost Ringers). Following a recovery period of 8 or more hours, and an 11-12 hour control flux period, the animals were infused with acid over a 5 minute period. After a one hour equilibration period, fish were either maintained in MDIBL seawater (~500 mM NaCl) or the external water was changed to 20% seawater (~100 mM NaCl; measured as Cl^-), or 4% seawater (~20 mM NaCl). During the control and post-infusion periods, water NH_4^+ and HCO_3^- were measured so that cumulative transfers of H^+ between the fish and the water could be calculated (Claiborne & Evans, ibid.). Likewise, blood samples (30-50 μl) were taken regularly throughout the experiment for the determination of plasma pH and total CO_2 and the calculation of plasma P_{CO_2} and $[HCO_3^-]$ (for details see Claiborne and Evans, ibid.).

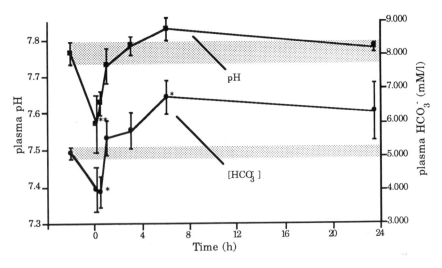

Figure 1. Plasma pH and $[HCO_3^-]$ in 5 seawater animals following acid infusion. Shaded bars represent pre-infusion control values. Infusion at hour 0. * = significant increase, ** = decrease, mean ± S.E.

Following acid infusion, sculpin in seawater exhibited a rapid decrease and then recovery of plasma pH and [HCO_3^-] which was complete within 1 hour post-infusion (Fig. 1). By hour 6, plasma pH had increased slightly (from 7.77 ± 0.03 to 7.83 ± 0.03) and [HCO_3^-] was ~32% higher than pre-infusion control (6.74 ± 0.50 versus 5.09 ± 0.18 mM; $p<0.05$, mean ± S.E., n=5). In animals exposed to 20% seawater following acid infusion, both plasma pH and [HCO_3^-] were well above control 7 hours post-infusion (Fig. 2, pH: 7.72 ± 0.05 → 7.90 ± 0.02, $p< 0.05$; [HCO_3^-]: 5.56 ± 0.04 → 7.80 ± 0.34 mM, $p<0.02$). 4% seawater induced a significant fall in plasma pH at hour 4 (Fig. 2; 7.77 ± 0.03 → 7.68 ± 0.01) and an extended decrease in [HCO_3^-] through hour 7 (5.03 ± 0.43 → 4.25 ± 0.39 mM, $p<0.02$).

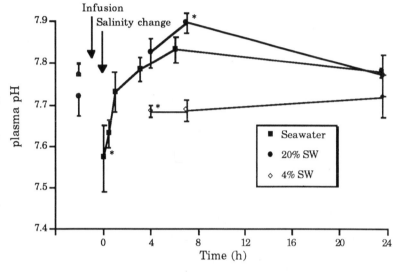

Figure 2. Plasma pH in fish exposed to seawater (n=5), 20% seawater (n=5), or 4% seawater (n=6) after acid infusion. Initial points are pre-infusion control values.

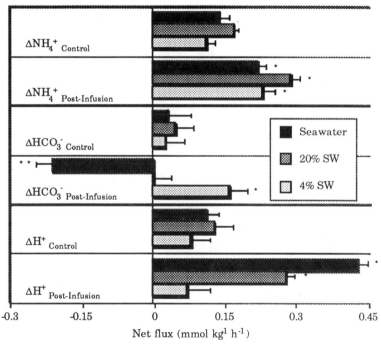

Figure 3. Net transfer rates of NH_4^+, HCO_3^-, and H^+ (mmol kg^{-1} h^{-1}) before (Control) and after (Post-Infusion) acid infusion in three groups of fish exposed to various salinities one hour after the infusion. Post-infusion flux calculated over 10.5 h for seawater group and 12 hours for 20% and 4% seawater series.

Figure 3 shows the net transfer rates of NH_4^+, HCO_3^-, and H^+ between the animal and the water measured over the pre-infusion control period and 10.5-13 hours post-infusion. In seawater, sculpin rapidly excreted 160% of the infused load. ΔH^+ efflux increased by 3.9 times (from 0.11 ±

0.03 to 0.43 ± 0.02 mmol kg^{-1} h^{-1}; p<0.001, n=5) in the first 10.5 hours following the infusion. The elevation in ΔH^+ was due to a significant increase in the rate of NH_4^+ loss and the net uptake of HCO_3^- (or excretion of H^+). Sculpin in 20% seawater excreted ~100% of the acid load in the first 12 hours as ΔH^+ increased from 0.13 ± 0.04 to 0.28 ± 0.04, mainly due to a 70% elevation in ΔNH_4^+ efflux. In contrast, ΔH^+ remained unchanged in animals exposed to 4% seawater, and they did not excrete the infused load. ΔNH_4^+ and ΔHCO_3^- increased in parallel fashion during this period (a net increase of ~12 mmol kg^{-1} h^{-1} over control rates; p<0.02), which produced little change in net H^+ loss. The patterns of efflux described for all experimental groups continued through hour 21.5-23.5 post-infusion.

Clearly, sculpin were able to rapidly compensate for the infused load when in seawater. Plasma pH and [HCO_3^-] were near normal within 1 hour, and were above controls by hour 6 (Fig. 1). The minimal and short-lived pH plasma depression immediately following the infusion was probably due to a slow uptake of the acid load from the intraperitoneal cavity (when compared to intravenous injection) and a rapid excretion of acid both branchially (~85%) and renally (~15%; calculated from Maren and Fine, ibid). Indeed, the fish exhibited an over-excretion of net H^+ to the water of 3.2 mmol kg^{-1} in the first 10.5 hours (Fig. 3), and the rate of excretion was still above control over the first 21.5 hours (resulting in a net H^+ loss of 4.3 mmol kg^{-1} when only 2 meq kg^{-1} had been infused). About 75% of the increase in ΔH^+ was due to a reversal of normal ΔHCO_3^- loss to a net uptake (or excretion of H^+, these are indistinguishable using the present methods; see Claiborne & Evans, ibid.). The remainder was driven by an elevation in total ammonia efflux. Similarly, a transbranchial over-compensation to acid infusion in the marine lemon sole (Parophrys vetulus; McDonald et al., J. Exp. Biol. 98:403-414, 1982) and the seawater adapted rainbow trout (Salmo gairdneri; Tang et al., J. exp. Biol. 134:297-312, 1988) have also been demonstrated. Thus, it appears that once the appropriate gill exchange mechanisms (see above) have been activated, net H^+ excretion continues well past the amount required for a compensation equivalent to the infused load.

When sculpin were acid loaded and subsequently exposed to decreased ambient salinities, the pattern of acid-base transfers changed. Fish in 20% seawater were able to regain normal (and above normal) plasma pH (Fig. 2) and [HCO_3^-], though the net acid excretion was mainly due to an increase in ΔNH_4^+ loss while ΔHCO_3^- was negligible (Fig. 3). This is supported by our finding that non-infused sculpin can also maintain near normal ΔH^+ transfers in 20% seawater (Claiborne & Perry, ibid.). In contrast, following exposure to 4% seawater, plasma pH and [HCO_3^-] remained below control for 4-7 hours, and the infused load was not excreted. A large net HCO_3^- loss at these low salinities (also observed in animals which were not acid loaded; Walton & Claiborne, ibid.) nullified an increase in net NH_4^+ excretion. We have hypothesized previously (Claiborne & Perry, ibid.) that changes in external [Na^+] may be responsible for the apparent HCO_3^- or H^+ transfer imbalances at these low salinities. The present data indicate that even when potential acid excretory mechanisms (eg., Na^+/H^+ exchange) should have been stimulated by increased internal H^+, low external [NaCl] may still limit the degree of acid-base compensation which can be achieved by these animals.

This study was funded by NSF DCM 86-02905 to JBC, and a Hearst Foundation Stipend to JBC and a Hearst Foundation Scholarship to EP.

RENAL RESPONSE TO SALINE INFUSION IN LEACH'S STORM PETREL (OCEANODROMA LEUCORHOA)

David L. Goldstein and Shawn Heflick
Department of Biological Sciences, Wright State University
Dayton, OH 45435

The avian kidney contains two general classes of nephrons. The reptilian-type (RT) nephrons are located superficially, have small glomeruli and short tubules, and lack loops of Henle; they are unable to produce a urine hyperosmotic to plasma. In contrast, the mammalian-type (MT) nephrons are located deeper in the kidney, have larger glomeruli and longer tubules, and do possess loops of Henle; they are responsible for the ability of birds to produce a concentrated urine. In the desert quail, both populations of nephrons function in urine production during conditions of hydration. However, under conditions that stimulate water conserving mechanisms (infusion of saline or antidiuretic hormone) the RT nephrons cease filtering and the MT nephrons, which are better able to conserve water, assume the preponderance of function. The balance of function between these nephron populations has never been examined in a bird that possesses a salt gland. In these species, ingested or infused sodium can be most efficiently excreted (with the least water loss) via the supra-orbital salt glands. The concentration of NaCl in fluids produced by these glands is essentially fixed, whereas the NaCl concentration of urine can vary. Hence, infusion of equal amounts of saline in different volumes of water should elicit a changing balance of function between the kidneys and salt glands. Based on the findings in the desert quail, one might expect that as the role of the kidney diminishes during times of high saline loading, one would observe concommitantly a shift in function among populations of nephrons. We tested this hypothesis using Leach's storm petrel, which produces the most concentrated salt gland fluids measured in a bird. We examined the renal response to saline infusion in chicks of this species, and examined qualitatively the relative roles of RT and MT nephrons under different levels of osmotic challenge.

We collected chicks of Leach's storm petrel, average age 33 $\pm$ 2 d, from nesting burrows on Little Duck Island. We kept birds in the laboratory for approximately one week prior to experimentation, during which time they were fed a blend of rehydrated krill, fresh fish, olive oil, and cod liver oil. Birds maintained body mass and wing growth was normal on this diet. Body mass at the time of experiments was 54 $\pm$ 1 g.

We evaluated kidney function using a constant infusion technique. During the experiments birds were placed facing into a darkened beaker; this simulated their natural nest environment, and they remained calm without any additional restraint or sedation. Each bird received, via a brachial vein, one of three saline infusions that provided the same NaCl load but in different volumes of water (0.55 M NaCl at 1.2 ml/h, 0.25 M NaCl at 2.6 ml/h, or 0.125 M NaCl at 5.3 ml/h). Infusion solutions also contained inulin (0.1%). A bird received its infusion for at least two hours, by which time urine flow and composition had stabilized for at least one hour. Blood was collected from the other brachial vein, and urine was collected via a cloacal cannula that was closed at one end to prevent contamination of urine by intestinal fluids.

Birds infused with the hyposmotic saline (125 mM) had urine flow rates of 4.4 $\pm$ 0.3 (mean $\pm$ SD) ml/min, 83% of the infused volume. Urine osmolality was 237 $\pm$ 10 mM/kg, of which approximately half (106 meq/l) was Na$^+$. Renal sodium excretion amounted to 71% of the infused load. GFR was 25.7 $\pm$ 5.8 ml/h in this group, and fractional water reabsorption was 84%. The GFR in birds receiving the 250 mM infusion (23.3 $\pm$ 4.6 ml/h) was not different from that in birds infused with 125 mM NaCl. However, fractional water reabsorption rose to 95.3%, and as a result urine flow was reduced to 0.9 $\pm$ 0.2 ml/h. Urine Na^+ rose to 252 meq/l, but urinary Na^+ excretion was just 35% of the infused load. In birds receiving the most concentrated infusion (550 mM NaCl), urine flow was markedly diminished. This resulted from both a large drop in GFR (to 3.8 $\pm$ 1.2 ml/h) and a further increase in fractional water reabsorption (to 97.7%). Urine osmolality was high (> 800 mM/kg), but only about 15% of the total concentration was Na^+ and less than 2% of the infused sodium load was excreted in the urine.

To evaluate relative roles of RT and MT nephrons, each bird received an infusion of Alcian blue for one half hour upon completion of the saline/inulin infusion. The infusion rate and NaCl concentrations were the same for the Alcian blue as for the saline/inulin infusions, and the concentrations of Alcian blue were adjusted so that all birds received the same total amount of stain. Alcian blue affixes to the fixed negative charges of the glomerular capillaries and filtration barrier, and so stains blue those glomeruli that are perfused and filtering. We measured the total number of stained glomeruli and the frequency distributions of their diameters to assess whether RT and/or MT nephrons were filtering.

The total number of stained glomeruli in birds receiving the 125 mM infusion was 59,022 $\pm$ 6,192. In view of the very high urine flow rate in these birds, it seems likely that all glomeruli were filtering, and therefore that the number of stained glomeruli in this group represents the total number of glomeruli in the petrel kidney (mean kidney mass was 0.45 $\pm$ 0.03 g/kidney). Birds receiving the 250 mM infusion had 55,030 $\pm$ 6,690 stained glomeruli, not different from the number in birds receiving 125 mM NaCl. In contrast, petrels receiving 550 mM NaCl had significantly fewer (37,176 $\pm$ 6990) stained glomeruli. In addition, the staining of the glomeruli in this latter group was markedly lighter than in the other two groups. This suggests that overall filtration was reduced during infusion of the most concentrated salt solution, and that some glomeruli ceased filtering altogether, at least within the limits of our detection. These data are consistent with those for GFR: no difference between the two lower concentrations of infusions, reduced filtration with the most concentrated infusion.

Given that the number of filtering glomeruli was reduced in the one group, we were interested in evaluating whether those nephrons that ceased filtering belonged to a particular size class. To assess this, we compared frequency distribution of diameters of stained glomeruli for the three groups of birds (Table 1).

Table 1. Percentages of stained glomeruli in different diameter (um) classes for petrels infused with three different solutions of NaCl (mean values, N = 6 for each group).

Diameter	125 mM	250 mM	550 mM
<50	4.5	1.7	1.1
50 - 65	31.5	28.7	12.3
65 - 80	44.7	40.9	37.0
80 - 95	17.3	19.1	30.3
95 - 110	5.3	7.2	11.8
>110	0.4	1.5	6.2

The overall range of sizes of stained glomeruli was similar in all three groups of birds. In addition, the frequency distributions of glomerular diameters were very similar for birds infused with 125 and 250 mM NaCl. However, in birds infused with the most concentrated solution there were diminished proportions of small glomeruli and greater proportions of larger ones. This suggests that reptilian-type nephrons, with their small glomeruli, selectively ceased filtering during salt loading. Again, the lighter staining of all glomeruli indicates that perfusion and filtration were reduced even in the larger mammalian-type nephrons under these conditions. This is consistent with the very low GFR's measured during infusion of 0.55 M NaCl. We cannot say whether the diminished staining represents a continuous perfusion and filtration at reduced rates, or instead intermittent function, with periods of filtration alternating with periods of cessation of filtration.

The high concentration of salt gland secretions in the petrel may be primarily related to the need for NaCl excretion related to the adult diet of marine invertebrates. However, the salt glands together with the ability to greatly diminish rates of glomerular filtration and urine flow also give the petrel chicks a very well developed ability to conserve water. This is likely important during the long nestling period, as the chicks are fed irregularly and sometimes as long intervals. Diminished function of the RT nephrons, with their inability to concentrate the urine, may help to prevent an unacceptable urinary water loss during such times.

This research was supported by a fellowship from the Lucille P. Markey Charitable Trust and by National Science Foundation grant DCB-8917616. Thanks to Vicki Reed and Jean Tipton for assisting with the morphological analyses.

EFFECTS OF pH, BARIUM AND COPPER ON INTESTINAL CHLORIDE TRANSPORT IN THE WINTER FLOUNDER (<u>PSEUDOPLEURONECTES AMERICANUS</u>)

Alan N. Charney and Angela Taglietta
Nephrology Section, VA Medical Center, NYU School of Medicine
New York, NY 10010

Decreases in extracellular pH (pHe) reduce the rate of Na and Cl absorption in the small intestine of the winter flounder. The major target of pH action is a Na-K-2Cl absorptive process located along the luminal membrane of intestinal epithelial cells. In addition, a poorly characterized Cl secretory process is stimulated by reductions in pHe. In a previous study, we found that amiloride at 1 mM also inhibited Na and Cl absorption suggesting that changes in pHi may mediate the effects of pHe. In the present study, we examined the possibility that pHe (or pHi) affects Na-K-2Cl co-transport indirectly by altering luminal membrane K conductance. In this scheme, reductions in pHi decrease K conductance and the resulting increase in intracellular K inhibits the Na-K-2Cl absorptive process. Such a mechanism has been observed in other epithelia with transport pathways similar to the flounder small intestine. To examine this possibility, we studied the effects of barium on Cl flux at concentrations known to inhibit K conductance in the flounder intestine. We also examined the link between the Cl absorptive and secretory processes by studying the effects of copper on Cl flux. In previous studies we found that this metal also inhibits Cl absorption.

Winter flounder were housed for 1-3 days before use. Small intestinal segments were stripped of serosa and studied under short circuit conditions in the Ussing chamber. Bathing solution pH was maintained at 7.8 by gassing a 20 mM bicarbonate teleost Ringers solution with 1% CO_2/99% O_2 (PCO_2 7 mmHg). Barium chloride was added to the mucosal or serosal bathing solution at a final concentration of 5 mM. In other experiments, cupric chloride was added to the mucosal or serosal solution at 20 or 100 uM.

	n	pHe	JClms	JClsm	JClnet	Isc
Control	9	7.8	9.3 ± 0.9	4.8 ± 0.9	4.5 ± 0.7	-1.8 ± 0.3
Barium 5mM (m)	9	7.8	8.6 ± 1.0	5.8 ± 1.1	2.8 ± 0.5	-0.9 ± 0.1
Copper 20uM (m)	4	7.8	8.1 ± 2.7	5.2 ± 2.0	2.9 ± 1.7	-0.1 ± 0.7
Copper 100uM (m)	10	7.8	6.4 ± 0.7	5.2 ± 1.0	1.2 ± 0.9	0.3 ± 0.3

(Results expressed as means$\pm$SE in $uEq/cm^2.h$)

In control tissues, net Cl absorption was present associated with a serosa negative PD. Mucosal barium inhibited Cl absorption by decreasing Jms and increasing Jsm. Furthermore, the presence of mucosal barium prevented the effect of a reduction in pHe on Cl flux (data not shown). Mucosal copper at concentrations between 20 uM and 100 uM also inhibited Cl absorption by decreasing Jms and increasing Jsm. The serosal addition of barium or copper did not affect Cl flux.

The results of the barium studies the possibility that pHe affects Cl absorption through an effect on luminal membrane K conductance. The similar effects of copper and pHe suggest that the Cl absorptive and secretory processes may be linked or coordinated in some way.

These studies were supported by the Dept. of Veterans Affairs and N.I.E.H.S. IP50 ES03828-06.

CHLORIDE SECRETION IN THE RECTAL GLAND OF <u>SQUALUS ACANTHIAS</u>: THE ROLE OF C-TYPE NATRIURETIC PEPTIDE (CNP)

R. Solomon[1], H. Solomon[2], D. Wolff[3], S. Hornung[4], H. Brignull[5], J. Landsberg[6], M. Silva[7], F.H. Epstein[8], P. Silva[1].

[1]Department of Medicine, Harvard Medical School, New England Deaconess Hospital, Joslin Diabetes Center, Boston, Massachusetts 02215
[2] Cambridge School of Weston, Weston, Massachusetts 02193
[3]Southwestern College, Winfield, KS 67156
[4] Siena College, Loudonville, New York 12211
[5] Bar Harbor High School, Bar Harbor, Maine 04609
[6] University of Southern California, Los Angeles, California 90007
[7] Mt. Desert Island Biological Laboratory, Salsbury Cove, ME 04672
[8]Department of Medicine, Harvard Medical School, Beth Israel Hospital, Boston, Massachusetts 02215

The rectal gland of Squalus acanthias can be stimulated by a variety of agonists including atrial natriuretic peptide (ANP) and vasoactive intestinal peptide (VIP). The present studies were conducted to evaluate the effects of other members of the family of natriuretic peptides. Using the homologous C-type natriuretic peptide (sCNP) for <u>Squalus acanthias</u>, we performed studies in the isolated perfused rectal gland to establish a physiologic role for this peptide. In addition, we separately perfused the glands with human C-type natriuretic peptide (hCNP), killifish CNP (kCNP), porcine BNP (pBNP) and human BNP (hBNP), human pre-pro ANP (hANP 31-67), and fragments and substitutions of the CNP peptide (hCNP 6-22, hCNP 7-21) to evaluate the requirements for ligand-receptor interaction.

Isolated shark rectal glands were perfused using a technique developed in our laboratory [Solomon, et al, Am. J. Physiol. 1985, 249; R348-R354]. The peptides hANP, hANP 31-67, and BNP were obtained from Peninsula Labs , killifish CNP (kCNP) was kindly supplied by Dr. D. Evans, and sCNP, hCNP, hCNP 6-22, hCNP 7-21, hCNP gly-9 were supplied by California Biotechnology Inc [Table 1]. Peptides were kept frozen until immediately prior to use when they were diluted into 1 ml of shark Ringer's solution. Following three baseline collection periods of 10 minutes each, the peptides were infused over 60 seconds into the perfusate line. Three additional 10 minute collections were then performed.

Table 1. Aminoacid structure of peptides studied (ring structure is outlined)

HUMAN ANP	S-L-R-R-S-S	C-F-G-G-R-M-D-R-I-G-A-Q-S-G-L-G-C	N-S-F-R-Y-COOH
PIG BNP	S-P-K-T-M-R-D-S-G	C-F-G-R-R-L-D-R-I-G-S-L-S-G-L-G-C	N-V-L-R-R-Y-COOH
HUMAN BNP	S-P-K-M-V-Q-G-S-G	C-F-C-R-K-M-D-R-I-S-S-S-S-G-L-G-C	K-V-L-R-R-H-COOH
HUMAN CNP	G-L-S-K-G	C-F-G-L-K-L-D-R-I-G-S-M-S-G-L-G-C	COOH
HUMAN CNP 6-2		C-F-G-L-K-L-D-R-I-G-S-M-S-G-L-G-C	COOH
HUMAN CNP 7-21		F-G-L-K-L-D-R-I-G-S-M-S-G-L-G	COOH
HUMAN CNP gly-9	G-L-S-K-G	C-F-G-G-K-L-D-R-I-G-S-M-S-G-L-G-C	COOH
KILLIFISH CNP	G-W-N-R-G	C-F-G-L-K-L-D-R-I-G-S-M-S-G-L-G-C	COOH
DOGFISH CNP	G-P-S-R-S	C-F-G-L-K-L-D-R-I-G-A-M-S-G-L-G-C	COOH

The effect of sCNP on chloride secretion is shown in Figure 1. The stimulatory effect seen was similar to that seen with VIP. A small stimulatory effect was first seen at a concentration of 10^{-9}M with maximal stimulation seen at a concentration of 10^{-6}M. Within the limits of the concentrations of sCNP examined half-maximal stimulation appears to be at 10^{-8}M. Figure 2 compares the effect of CNP's from different species on chloride secretion by the rectal gland. The effect of sCNP is similar to that of kCNP and an order of magnitude greater than that of hCNP. Figure 3 shows the effect of various modifications of the CNP molecule. Removal of the aminoterminal end (hCNP 6-22) causes a significant reduction in the stimulatory effect. Substitution of the leucine in position 9 by a glycine did not significantly alter the functional effect of the molecule (hCNP gly-9 vs. hCNP). Surprisingly the ringless structure (hCNP 7-21) made up of aminoacids 7 to 21 had a functional effect that was only an order of magnitude lower than that of the ring alone. Figure 4 compares the effect of sCNP with that of hANP, pork and human BNP and hANP 31-67. As can be seen the effect of hANP is considerably smaller than that of sCNP and that of pBNP is even more so. hANP 31-67 had virtually no effect.

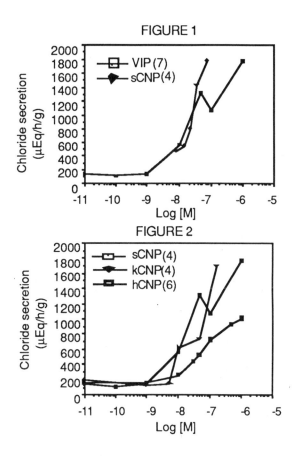

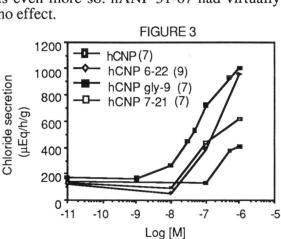

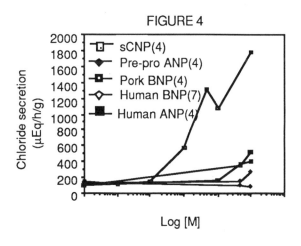

These results support a physiologic effect of sCNP to stimulate salt secretion by the shark rectal gland. sCNP is a 22 aminoacid peptide containing a 17 member ring on the COOH-terminal end. kCNP and sCNP differ in three aminoacids in the NH_2-terminal end, positions 2,3, and 5, and in position 16 in the ring structure. Both peptides had similar dose response curves suggesting that these amino acid differences were not important. On the other hand, hCNP differs from sCNP at positions 2,4,5 and 16 and from kCNP at positions 2,3,4 and is one order of magnitude less potent than sCNP or kCNP. This suggests that position 4 in the NH_2-terminal end of the molecule is of functional significance. Further support for the importance of the NH_2-terminal end of the molecule is suggested by the marked reduction in response to hCNP 6-22 containing the ring structure alone compared to intact hCNP. In addition to the importance of the NH_2-terminal end of the molecule, the ring may also be of some functional significance. The linear fragment, hCNP 7-21, and the intact ring, hCNP 6-22, were still capable of stimulating chloride secretion albeit at a markedly blunted maximal levels and only at very high concentrations ($>5 \times 10^{-7}M$ and $>10^{-7}M$) of peptides respectively. In bovine aorta smooth muscle cells which contain the ANPR-B receptor subtype, hCNP and hCNP 6-22 have similar dose response curves for the generation of cGMP. The linear fragment, hCNP 7-21, on the other hand, is inactive (A. Protter, California Biotechnology, Inc, personal communication). The substitution of glycine for leucine at position 9 in the ring also reduced the stimulatory effect (hCNP gly-9 vs hCNP) suggesting a functional role for this site as well. In the bovine aorta smooth muscle preparation, hCNP gly-9 is also inactive (Ibid.).

BNP interacts primarily with the ANPR-A receptor. In the isolated perfused rectal gland it has a dose response curve shifted well to the right of that of hANP. Taken together with the results obtained with the various CNP's, we suggest that the biologically active receptor for sCNP in shark rectal gland resembles the ANPR-B receptor. In addition, these studies suggest that sCNP may be a physiologic regulator of rectal gland function.

Supported by grant NIH DK 18078 (FHE) and the Pew (FHE,PS), W.R. Hearst Foundation (FHE) and Burroughs-Welcome Foundation grants to MDIBL.

EFFECT OF PROTEIN KINASE C ACTIVATION ON CHLORIDE SECRETION BY THE RECTAL GLAND OF SQUALUS ACANTHIAS

Patricio Silva,[1] Richard Solomon[1], Heather Brignull, Sonya Hornung, Judd Landsberg, Hadley Solomon, Douglas Wolff, and Franklin H. Epstein[2]
Department of Medicine, Harvard Medical School and New England Deaconess Hospital and Joslin Diabetes Center[1] and Beth Israel Hospital,[2] Boston, MA 02215

In previously unreported experiments we identified inositol mono-, di- and triphosphate in the shark rectal gland indicating that this pathway is present in that epithelium. Recently, Ecay et al., showed that vasoactive intestinal peptide (VIP) stimulates inositol phosphate release in cultured rectal gland cells grown in suspension (Bull MDIBL 28:72-73, 1989). Brand et al., confirmed that VIP activates the inositol phosphate pathway showing that it activates phospholipase C in plasma membranes of shark rectal gland. These authors also showed that 2-chloro adenosine also activated phospholipase C in the same preparation but that human alpha-atrial natriuretic peptide (ANP) failed to do so (Bull MDIBL 29:96-97, 1990). Simultaneously, Ecay et al., found that rat ANP I and III increased the levels of inositol mono- and diphosphates in cultured rectal gland tubules (J Cell Physiol 146:407-16, 1991).These authors also found that ionomycin 10^{-6}M increased total inositol phosphates.More recently, Feero et al., have found that phorbol myristate acetate (PMA) 10^{-7}M increased short circuit current in cultured rectal gland cells grown to confluence, an effect not reproduced by an inactive phorbol ester and decreased by the inhibitor of protein kinase C, staurosporine (Bull MDIBL 30:63-64, 1991). The same group has also reported that ionomycin 10^{-6}M stimulates short circuit current in the same preparation. To date there is no reported data on the role of the inositol phosphate pathway on chloride secretion in the intact rectal gland. The present experiments were performed to investigate its role in isolated perfused, but otherwise intact, rectal glands.

Isolated shark rectal glands were perfused using a technique developed in our laboratory. Dogfish were pithed and the rectal glands removed by an abdominal incision. The rectal gland artery, vein and duct were catheterized and the glands placed in a glass perfusion chamber maintained at a temperature of 15° C with running sea water. The glands were perfused by gravity at a pressure of 40 mm Hg. The composition of the perfusate was (in mM): Na, 280; Cl, 280; K, 5; bicarbonate, 8; phosphate, 1; Ca, 2.5; Mg, 1; sulfate, 0.5; urea, 350; glucose, 5; ph, 7.6 when gassed with 99% O_2/ 1% CO_2. Rectal gland secretion was collected in tared 1.5 ml centrifuge tubes over 10 minute intervals. Chloride concentration in the rectal gland secretion was measured by amperometric titration. Glands were stimulated to secrete chloride with vasoactive intestinal peptide (VIP) 1.5 x 10^{-9}M dibutyryl cyclic AMP 5 x 10^{-5}M and theophylline 2.5 x 10^{-4}M.

In two experiments, TPA at a concentration of 1.6 x 10^{-6}M had no effect on chloride secretion in isolated perfused rectal glands stimulated to secrete chloride with VIP, Figure 1. Oleyl acetyl glycerol at a concentration of 6.3 x 10^{-5}M also had no effect on VIP stimulated chloride secretion as also shown in Figure 1.

Both TPA and oleyl acetyl glycerol activate protein kinase C but do not increase the calcium concentration of the cytoplasm that is also required for the activation of protein kinase C. Therefore, experiments were done in which TPA was used in combination with ionomycin that increases the entry of calcium into the cell. TPA, 1.6 x 10^{-6}M, and ionomycin 10^{-6}M had no effect on chloride secretion, Figure 2.

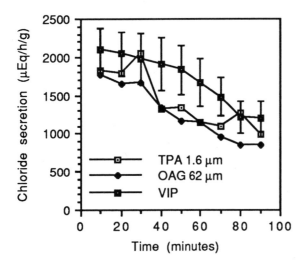

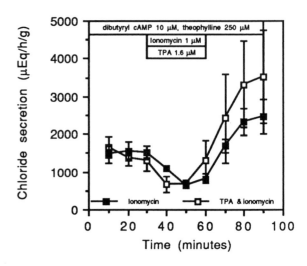

Figure 1. Effect of TPA 1.6 μM and oleyl acetyl glycerol 62 μM on chloride secretion in rectal glands stimulated with VIP. Neither TPA nor OAG had any effect on chloride secretion. Values are mean ± SEM, n: VIP=6, TPA=2, OAG=2.

Figure 2. Effect of TPA 1.6 μM combined with ionomycin 1 μM and ionomycin 1 μM alone on chloride secretion in glands stimulated with theophylline 2.5×10^{-4}M and dibutyryl cyclic AMP 5×10^{-5}M Values are mean ± SEM, n: TPA plus ionmycin=6, ionomycin alone=4.

In glands stimulated with theophylline 2.5×10^{-4}M and dibutyryl cyclic AMP 5×10^{-5}M the combination of TPA and ionomycin caused a small but statistically not significant decrease in chloride secretion followed by a large increase in chloride secretion.

The increase in chloride secretion was the result of a large increase in fluidsecretion with a fall in chloride concentration. The increase in chloride secretion started after the TPA perfusion had been stopped and persisted for the remainder of the experiment. Ionomycin 10^{-6}M alone produced the same effect as that seen in combination with TPA.

Figure 3. Effect of ionomycin 10^{-7}M followed by TPA 1 μM on chloride secretion in glands stimulated with theophylline 2.5×10^{-4}M and dibutyryl cyclic AMP 5×10^{-5}M. At this concentration ionomycin had no effect on clhloride secretion nor did TPA. Values are mean ± SEM, n=6.

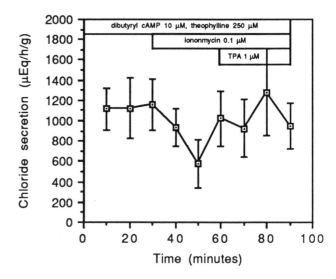

Lowering the concentration of ionomycin 10 fold to 10^{-7}M prevented all effects of ionomycin. TPA added after the ionomycin, while the latter was still in the perfusate, had no additional effect as can be seen in Figure 3.

Ionomycin 10^{-6}M evoked the same secretory effect associated with a decrease in the concentration of chloride in glands that were not stimulated to secrete chloride. TPA, however, had no effect on unstimulated glands, Figure 4..

The above experiments show that activation of protein kinase C has no effect on chloride secretion by the rectal gland. Neither TPA nor oleyl acetyl glycerol, that activate protein kinase C, had any effect on chloride secretion in the perfused rectal gland when used alone. When the calcium ionophore, ionomycin was used at a concentration of $10^{-7}M$ in addition to TPA again there was no effect. Ionomycin used alone at a concentration of $10^{-7}M$ had no effect, but at a concentration of $10^{-6}M$ it produced a large increase in fluid

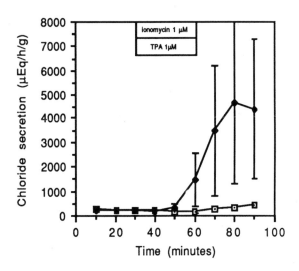

Figure 4. Effect of TPA 1 μM or ionomycin 1 μM on chloride secretion in unstimulated glands. Values are mean ± SEM, n: TPA (open squares)=6, ionomycin (closed diamonds)=4.

secretion associated with a fall in the concentration of chloride in the secretory fluid. The fall in the concentration of chloride suggests that there was ionomycin-induced damage to the rectal gland epithelia. The concentration of chloride in the rectal gland secretion remains constant at a level approximately twice that of the perfusate when chloride secretion is stimulated. This is the result of the luminal membrane of the rectal gland cells impermeability to urea. If the cell membrane becomes permeable to urea, or urea is removed from the perfusate, the concentration of chloride in the perfusate falls and the volume of fluid increases. The finding that ionomycin increases chloride secretion while decreasing the concentration of chloride suggests that it is increasing luminal permeability.

Supported by grants NIH DK 18078, and NIEHS ESO 3828 and the Pew, W.R. Hearst Foundation and Burroughs-Welcome Foundation grants to Heather Brignull, Sonya Hornung, and Douglas Wolff.

EFFECT OF MERCURY ON CHLORIDE SECRETION BY THE RECTAL GLAND OF <u>SQUALUS ACANTHIAS</u>

Patricio Silva,[1] Richard Solomon[1] Heather Brignull, Sonya Hornung, Judd
Landsberg, Hadley Solomon, Douglas Wolff, and Frnaklin H. Epstein[2]
Departments of Medicine, Harvard Medical School and New England Deaconess
Hospital and Joslin Diabetes Center[1] and Beth Israel Hospital,[2] Boston, MA
02215

We have previously examined the effect of organic and inorganic
mercurials on chloride secretion by the rectal gland. We found that mercuric
chloride inhibits chloride secretion in a dose dependent and irreversible way.
Mersalyl, an organic mercurial had no effect. This findings are interesting
because they are contrary to those observed in the mammalian kidney where
their site of action, the thick ascending limb, is thought to be the mammalian
counterpart to the rectal gland. In this report we explore the effect of
additional organic mercurials and also investigate the effects of
dithiotreitol (DTT) a protector of sulfhydryl groups. We also examined the
effect of cadmium on the inhibitory effect of mercuric chloride because
cadmium prevents the effect of mercury in some tissues (Webb et al. Chem Biol
Interact 14:357-69, 1976).

Isolated shark rectal glands were perfused using a technique developed
in our laboratory. Dogfish were pithed and the rectal glands removed by an
abdominal incision. The rectal gland artery, vein and duct were catheterized
and the glands placed in a glass perfusion chamber maintained at a temperature
of 15° C with running sea water. The glands were perfused by gravity at a
pressure of 40 mm Hg. The composition of the perfusate was (in mM): Na, 280;
Cl, 280; K, 5; bicarbonate, 8; phosphate, 1; Ca, 2.5; Mg, 1; sulfate, 0.5;
urea, 350; glucose, 5; ph, 7.6 when gassed with 99% O_2/ 1% CO_2. Rectal gland
secretion was collected in tared 1.5 ml centrifuge tubes over 10 minute
intervals. Chloride concentration in the rectal gland secretion was measured
by amperometric titration. Glands were stimulated to secrete chloride with
dibutyryl cyclic AMP 5 x 10^{-5}M and theophylline 2.5 x 10^{-4}M.

Effect of organic mercurials on chloride secretion

Figure 1 shows that meralluride at
a concentration of 10^{-4}M had no effect
on chloride secretion. In the initial
experiments meralluride was dissolved in
DMSO. Because of the possibility that
the drug was not adequately dissolved we
repeated the experiments with
meralluride dissolved in glacial acetic
acid and the perfusate solution titrated
back to 7.6 after the addition of the
meralluride. Again there was no effect.
The results of both series of
experiments were pooled together.

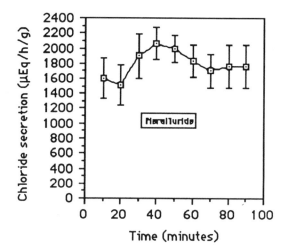

Figure 1. Effect of meralluride 10^{-4}M
on chloride secretion. Meralluride was
added to the perfusate during the time
indicated by the box. There was no
effect of meralluride on the secretion
of chloride. Values are mean ± SEM, n=6.

Another organic mercurial, PCMBS was then tested. Figure 2 shows a dose response curve for PCMBS. There was no effect on chloride secretion. A small but statistically not significant effect was discernible at 10^{-4}M.

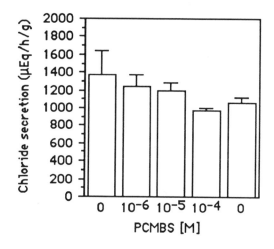

Figure 2. Effect of PCMBS on chloride secretion. Each column represents the sequential addition of increasing concentrations of PCMBS after an initial control period and ended with another control period. All periods were 30 minutes in duration There was no effect of PCMBS on the secretion of chloride by the rectal gland. Values are mean ± SEM, n=6.

Effect of DTT

DTT is known to prevent the toxic effect of mercuric chloride in many cell systems. We therefore examined the effect of DTT on the inhibitory effect of mercuric chloride. Figure 3 shows that DTT reduced the effect of mercuric chloride. In addition, the effect of mercuric chloride was no longer irreversible in the presence of DTT.

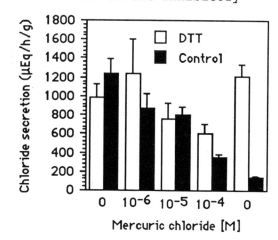

Figure 3. Effect of DTT on the toxic effect of mercuric chloride. DTT reduced the effect of mercuric chloride seen at 10^{-6}M and also at 10^{-4}M. In the presence of DTT the effect of mercuric chloride was no longer irreversible, compare DTT with control in the last control period labeled 0. Values are mean ± SEM, n=11 for DTT experiments and n=8 for the control.

Effect of cadmium

Because cadmium has been found to antagonize the effects of inorganic mercury we tested the effect of cadmium chloride on the toxic effect of mercuric chloride. Figure 4 summarizes the findings. Cadmium chloride did not prevent the effect of mercuric chloride.

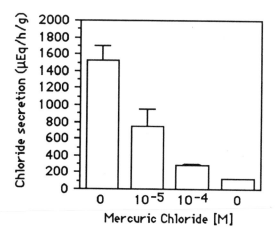

Figure 4. Effect of cadmium chloride on the toxic effect of mercuric chloride. Cadmium at a concentration 250 mM did not prevent the toxic effect of mercury. Values are mean ± SEM, n=3.

These experiments complement and corroborate the results we had previously obtained. Inorganic mercury has a toxic effect on the secretion of chloride by the rectal gland. The effect of mercury appears to be linked to its ability to bind to sulfhydryl groups inasmuch as it is prevented by DTT that protects these groups, although it is also possible that DTT may be acting as a sink and preventing mercury from reaching the tissue. A consistent and puzzling result is the lack of effect of organic mercurials. Neither mersalyl, previously reported to be without effect, nor meralluride or PCMBS had an inhibitory effect on chloride secretion at concentration as high as 10^{-4}M. Of note is the observation by Kleinzeller et al. (Biochim. Biophys. Acta 1025:21-31, 1990) that PCMBS and other organic mercurials induce swelling of slices of shark rectal gland. Our results are particularly interesting because they are opposite to those observed in the mammalian kidney. In the kidney mercurials exert their effect on the thick ascending limb of the loop of Henle, a segment of the nephron that is considered in general terms to be homologous to the rectal gland. In the kidney, organic mercurials like mersalyl and meralluride have a powerful effect while inorganic mercurials are not nearly as effective. Moreover, in isolated perfused thick ascending limbs of the loop of Henle, mersalyl inhibits chloride reabsorption. Interestingly, this effect is prevented by PCMB. We have no explanation for these differences. The effect of organic mercurials on the kidney is thought to depend on their ability to release inorganic mercury. Both mersalyl and meralluride are very efficient diuretics in the kidney whereas they have no effect in the rectal gland. Another explanation for the differential effect of inorganic versus organic mercurials is that their effect depends on pH. We tested for this in previous experiments and found that reducing the pH did not evoke an inhibitory effect of mersalyl. On the other hand, inorganic mercury that is not nearly as efficient in the kidney has a clear effect in the rectal gland. We conclude from this experiments that the site of action of mercurials in the thick ascending limb of the mammalian kidney is not present or not accessible in the rectal gland. Alternatively, if the release of inorganic mercury from the organic compounds is the necessary step for an effect of organic mercurials, the rectal glands may lack the capacity to do so thus rendering these compounds completely ineffective.

Supported by grants NIH DK 18078, and NIEHS ESO 3828 and the Pew, W.R. Hearst Foundation and Burroughs-Welcome Foundation grants to Heather Brignull, Sonya Hornung, and Douglas Wolff.

SYNTHETIC SHARK CNP BASED ON THE AMINO ACID SEQUENCE OF CLONED PRE-PRO SHARK HEART CNP POTENTLY STIMULATES CHLORIDE SECRETION IN THE PERFUSED SHARK RECTAL GLAND

J.N. Forrest, Jr.,[1] G.G. Kelley, [1] J.K. Forrest,[1] D. Opdyke,[1] J.P. Schofield,[2] and C. Aller[1]

[1]Department of Medicine,Yale University School of Medicine New Haven, CT 06510 and [2]MRC Molecular Genetics Unit, Hills Rd, Cambridge, England

Previous studies have suggested that cardiac ANP-like peptides play an important role in sodium chloride secretion by the rectal gland. Investigation of the synthesis, regulation, sites of action and signal transduction mechanisms of ANP in the rectal gland have been limited because native cardiac peptides have not been identified in the shark.

We recently cloned and sequenced pre-pro C-type natriuretic peptide (CNP) from the shark heart and identified a primary amino acid sequence that is distinctly different from all other cardiac peptides known to date (Schofield et al., Am J. Physiol. 30:F734-F739, 1991). This was the first report of a cDNA encoding CNP specifically in cardiac tissue and indeed in any non-neuronal tissue. This study also indicated that the mRNA encoding shark heart CNP is a highly abundant message. We have now synthesized shark heart CNP based on the deduced amino acid sequence of cloned CNP and herein report a comparison of the potency of shark heart CNP vs. mammalian ANP in the stimulation of chloride secretions in the in vitro perfused shark rectal gland.

Shark heart CNP was prepared at the Yale Protein and Nucleic Acid Chemistry Facility by the solid-phase method using an Applied Biosystems 430A peptide synthesizer (Foster City, CA). Standard tBOC-amino acids were incorporated as their hydroxybenzotriazole) activated esters in dimethylformamide. The peptide was cleaved using hydrogen fluoride and the resulting crude material, containing the reduced peptide, was purified by reverse-phase HPLC on Vydac C18 columns. Elution of the peptide using an acetonitrile gradient in 0.05% trifluoroacetic acid gave purified, reduced peptide according to amino acid analysis, analytical HPLC, and FAB-mass spectroscopy. The peptide was allowed to form an intramolecular disulfide by adjusting the pH of the reduced peptide to 7.5 with potassium phosphate buffer and allowing the solution to stir, uncovered overnight. The oxidized peptide was repurified as described above, pooled, and lyophilized. 25 mg of dried, oxidized material was obtained and characterized as greater than 90% pure by amino acid analysis, analytical HPLC and FAB-mass spectroscopy.

Figure 1 demonstrates the effect of shark CNP (10 nM) added to the perfusate of the in vitro perfused rectal gland after thirty minutes of basal perfusion.

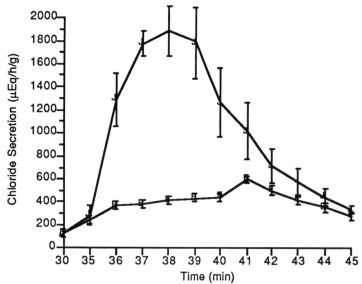

Figure 1. Response to 10 nM shark CNP (upper curve) vs 10 nM rat ANP (lower curve) on chloride secretion (µEq/h/g) in the perfused shark rectal gland.

As shown in Fig 1 (left), 10 nM shark CNP produces a profound stimulation of chloride secretion compared to 10 nM ANP. This stimulation reached a peak 8-9 min after addition and declined thereafter. In other experiments in the perfused gland shark CNP produces a marked increase in chloride secretion at each concentration studied (1-100 nM) in comparison to mammalian ANP. CNP (10 nM) produced a striking elevation in tissue cyclic GMP compared to controls. Homologous shark CNP thus has marked stimulatory effects on chloride

secretion in comparison to mammalian ANP at each concentration studied. The amino acid sequence of shark heart CNP compared to ANP and BNP is given below. Identical amino acids compared to shark CNP are indicated by letters.

Shark CNP Structure Compared to ANP and BNP

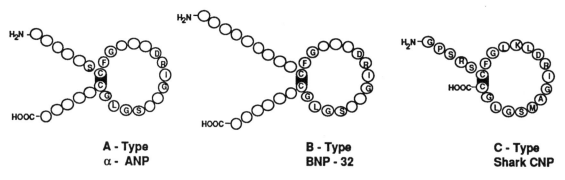

| A - Type | B - Type | C - Type |
| α - ANP | BNP - 32 | Shark CNP |

Prior to our identification of CNP in the shark heart, CNP had been identified only in brain tissue of five species (pig, rat, frog, killifish and eel) (Schofield et al., Am J. Physiol. 30:F734-F739, 1991). It is intriguing that none of the brain CNPs contain a serine residue in position 2 as found in shark CNP and mammalian ANP. With the availability of this first cardiac natriuretic peptide from the shark we will be able to carry out structural-functional studies and plasma measurements to define the synthesis, regulation, sites of action and signal transduction mechanisms of CNP in the regulation of chloride transport in the shark.

This work was supported in part by NIH DK 34208 (Dr Forrest) and NIH EHS-P30-ES03828 (Membrane Toxicity Studies).

cAMP-STIMULATED CHLORIDE CURRENT IN <u>XENOPUS</u> OOCYTES EXPRESSING mRNA FROM THE ALKALINE GLAND OF THE LITTLE SKATE <u>RAJA ERINACEA</u>

Roger T. Worrell[1,2], W. Gregory Feero[3], Sonia A. Cunningham[1],
Marybeth Howard[1], David C. Dawson[4], and Raymond A. Frizzell[1]
[1]Dept. Physiol. & Biophys., Univ. of Alabama at B'ham, Birmingham, AL 35294
[2]Current address: Dept. Physiol., Emory Univ. Sch. of Med., Atlanta, GA 30322
[3]School of Medicine, Univ. of Pittsburgh, Pittsburgh, PA 15261
[4]Dept. Physiology, Univ. of Michigan, Ann Arbor, MI 48104

The alkaline gland of the male skate secretes a fluid with a pH >9. This fluid presumably neutralizes the acid urine and preserves sperm viability (Maren et al., Comp. Biochem. Physiol. 10:1, 1963). More recent studies by Smith (AJP 248:R346, 1985) employed both transepithelial flux studies and microelectrode measurements which identified active chloride secretion as the principle electrolyte transport event. Therefore, the cellular mechanism of bicarbonate secretion would require a chloride bicarbonate exchange mechanism in parallel with the apical chloride conductance to account for luminal alkalinalization. The purpose of this study was to determine whether this apical membrane conductance could be expressed in <u>Xenopus</u> oocytes, stimulated by cAMP, and whether it is likely to arise from the expression by alkaline gland cells of a protein homologous to the cystic fibrosis transmembrane conductance regulator (CFTR).

The bilobular glands from two skates were excised, and the epithelial layer scraped free with a glass slide, yielding approximately 0.5 g tissue wet weight. PolyA$^+$ mRNA from alkaline gland cells was isolated by poly(dT) chromatography using the FastTrack kit of Invitrogen. A final spin at 500 X g was included to eliminate particulate matter that would interfere with mRNA injection. 50 ng mRNA was injected into <u>Xenopus laevis</u> oocytes and 2-3 days later the transmembrane currents were recorded using double-electrode voltage clamp. Methods are described in more detail by Cunningham et al. (AJP:Cell, 1992, in press).

Currents in mRNA-injected oocytes were initially low (<0.2 μA at ±80 mV) and were similar in magnitude to those observed in uninjected cells. Stimulation of mRNA-injected oocytes with a cAMP cocktail containing 10 μM forskolin, 200 μM 8-(4-chlorophenylthio)-adenosine (3':5'-cyclic monophosphate) (cpt-cAMP), 1 mM 3-isobutyl-1-methyl xanthine (IBMX) produced a 5-7 fold increase in transmembrane currents at ±80mV (n=6). This stimulation was reversible on removal of the cAMP cocktail. Figure 1 illustrates a representative current-voltage (I-V) relation showing the basal and stimulated currents, and Table I summarizes the cAMP-induced changes in membrane potential (V_m) and conductance (G_m) associated with this stimulation. As indicated in Fig. 1, the effects of cAMP were entirely reversible. Currents were usually elevated to a steady level within 2-5 min, and returned to control values approximately 10-15 min following cAMP cocktail washout from the bath. During current stimulation, reduction of bath chloride concentration shifted the I-V relation in the direction expected for stimulation of a chloride conductance pathway (Fig.1). Removal of chloride produced a 14 mV depolarization of the membrane potential and a reduction in membrane conductance (Table I). This represents a 22% inhibition of membrane conductance when bath chloride was reduced from 96 to 13 mM. Much of the non-chloride dependent conductance associated with cAMP stimulation was blocked by addition of 2 mM Ba. A similar phenomenon was associated with the expression of human CFTR mRNA in this system which may be due to either a direct or indirect effect of CFTR on membrane potassium conductance (Cunningham et al., AJP:Cell 1992, in press).

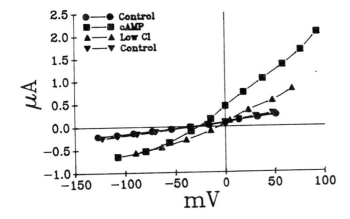

Figure 1. Current-voltage relations of an oocyte injected with 50 ng skate alkaline gland mRNA in the presence and absence (control) of cAMP cocktail or cAMP cocktail + low Cl media (13mM). Currents represent an average of the values obtained from 100 to 200 ms during a 250 ms voltage pulse.

Table 1. Effect of skate alkaline gland mRNA expression on oocyte membrane potential (V_m) and conductance (G_m). Values are mean $\pm$ SE (n).

	E_m (mV)	G_m (mV)
Control	-33$\pm$4 (6)	2.3$\pm$0.4 (6)
cAMP	-25$\pm$3 (6)	10.4$\pm$1.3 (6)
Low Cl	-11$\pm$4 (6)	8.4$\pm$1.3 (4)

Northern blot analysis of the polyA$^+$ mRNA isolated from skate alkaline gland was carried out using [^{32}P]-labeled CFTR cDNA isolated from shark rectal gland, a model chloride secreting epithelium (CFTR cDNA was kindly provided by Dr. J. Riordan, Univ. of Toronto). The preliminary results from mRNA blotting suggest that the alkaline gland expresses an elasmobranch homolog of CFTR, which is therefore a reasonable candidate for mediating the cAMP response.

Our findings are consistent with the following conclusions: 1) The expression of skate alkaline gland mRNA in Xenopus oocytes produces a cAMP-stimulated conductance response which is similar to that observed from expression of human CFTR. This is consistent with the idea that a cAMP-regulated chloride conductance in the apical membrane of alkaline gland cells provides the driving force underlying bicarbonate secretion and luminal alkalization. 2) The currents induced in oocytes expressing alkaline gland mRNA are similar to those observed with shark rectal gland mRNA (Sullivan et al., AJP, 260:C664, 1991), and presumably arise from expression of elasmobranch CFTR homologs in both tissues. 3) Skate alkaline gland expresses message that hybridizes with a probe derived from shark rectal gland, suggesting that the species which lie in the same subclass elasmobranchii express homologous CFTR-like proteins. 4) The cAMP- stimulated chloride conductance which is similar to that derived from rectal gland, therefore, probably results from CFTR itself. Thus, CFTR plays a role both in chloride secreting epithelia, such as rectal gland, and by coupling to other transport processes, this conductance can form the basis for bicarbonate secretion.

Supported by grants from the NIH (DK38518 to RAF & DK29786 to DCD), NIEHS Biomedical Sciences Center Grant, and the Cystic Fibrosis Foundation.

NATRIURETIC PEPTIDE (CNP) STIMULATION OF CHLORIDE CONDUCTANCE IN XENOPUS OOCYTES EXPRESSING mRNA FROM RECTAL GLAND OF THE DOGFISH SHARK SQUALUS ACANTHIAS

Roger T. Worrell[1,2], W. Gregory Feero[3], David C. Dawson[4],
and Raymond A. Frizzell[1]
[1]Dept. Physiol. & Biophys., Univ. of Alabama at B'ham, Birmingham, AL 35294
[2]Current address: Dept. Physiol., Emory Univ. Sch. of Med., Atlanta, GA 30322
[3]School of Medicine, Univ. of Pittsburgh, Pittsburgh, PA 15261
[4]Dept. Physiology, Univ. of Michigan, Ann Arbor, MI 48104

The family of natriuretic peptides derived from atrial extracts includes CNP, a major natriuretic peptide found in the brain of bony fish (Evans, D.H., Rev. Physiol. 52:43, 1990). CNP and its homologues are thought to serve an osmoregulatory role in fish affecting both absorptive and secretory salt transport processes. In perfused shark rectal gland, atrial peptides are thought to act secondary to stimulation of release of the neuropeptide VIP (vasoactive intestinal peptide) from nerve terminals (Silva et al., AJP, 252:F99, 1987). Other studies indicate that natriuretic peptides can stimulate chloride secretion from shark rectal gland cells in primary culture, and the stimulation is thought to involve an increase in cellular cGMP (Karnaky et al., MDIBL Bul., 29:86,1990). Although the cultured cell studies suggest that natriuretic peptides can have a direct effect on chloride secretory cells, they do not rule out the possibility that a paracrine stimulation of secretion exists in the cultures due to contamination by other cell types. We approached the question of direct versus indirect stimulation by expressing shark rectal gland mRNA in Xenopus oocytes with the assumption that the oocyte would express both receptor-mediated and second messenger-activated conductance properties of the epithelial cells.

The methods are similar to those described in the accompanying paper Worrell et al. (this issue). For these experiments, polyA$^+$ mRNA was derived from freshly isolated shark rectal gland. Oocytes were injected with 50 ng mRNA and oocyte current assayed 2-3 days later.

Figure 1 shows that the basal currents were low (<0.2 μA) at ±80 mV relative to the resting membrane potential. Addition a cAMP cocktail containing 10 μM forskolin, 200 μM 8-(4-chlorophenylthio)-adenosine (3':5'-cyclic monophosphate) (cpt-cAMP), and 1 mM 3-isobutyl-1-methyl xanthine (IBMX) stimulated a marked elevation of membrane currents, a depolarization of the resting membrane potential, and a large increase in the membrane conductance. The stimulated currents were chloride selective since reducing bath chloride from 96 to 13 mM shifted the current-voltage (I-V) relation to the right as anticipated from the change in chloride equilibrium potential and reduced the conductance. These effects were entirely reversible. Similar findings have been reported by Sullivan et al. (AJP, 260:C664, 1991) in oocytes expressing rectal gland mRNA. As illustrated in Figure 2, current was also stimulated by bath addition of 10^{-8} M CNP derived from killifish brain (CNP was kindly provided by Drs. D. Evans and K.J. Karnaky). Activation by CNP was also reversible (Fig. 2), and the CNP-stimulated currents shifted as anticipated for stimulation of a chloride conductance pathway in a manner similar to that shown in Fig.1 (data not shown).

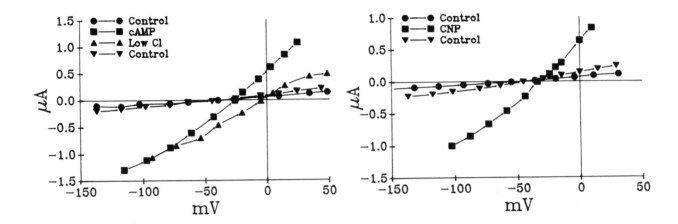

Figure 1. Current-voltage relations of an oocyte injected with 50 ng shark rectal gland mRNA in the presence and absence (control) of cAMP cocktail or cAMP cocktail + low Cl media (13mM). Currents represent an average of the values obtained from 100 to 200 ms during a 250 ms voltage pulse.

Figure 2. Current-voltage relations of an oocyte injected with 50 ng shark rectal gland mRNA in the presence or absence (control) of 10^{-8} M killifish CNP. Current determination was as in Fig. 1.

These findings are consistent with the idea that natriuretic peptides can act directly on shark rectal gland epithelial cells to stimulate chloride secretion since the effect of CNP on oocytes expressing shark rectal mRNA cannot involve a paracrine (cell-to-cell) pathway. The currents stimulated by CNP are similar to those arising from cAMP stimulation. However, the CNP effect may involve an increase in oocyte cGMP level since increased cGMP in response to CNP has been shown to occur in isolated shark rectal gland cells (Karnaky et al., MDIBL Bul., 29:86, 1990). Thus, there are probably two mechanisms by which natriuretic peptides stimulate chloride secretion; one involving a direct stimulation of the epithelial cells by CNP and another involving VIP release from CNP stimulated peritubular nerve terminals.

Supported by grants from the NIH (DK35815 to RAF & DK29786 to DCD), NIEHS Biomedical Science Center Grant, and the Cystic Fibrosis Foundation.

EVIDENCE FOR A Na^+/H^+ EXCHANGER IN THE CHLORIDE CELLS OF <u>FUNDULUS HETEROCLITUS</u>

Jose A. Zaduniasky[1], and Benjamin Lowenstein[2]
[1]Department of Physiology and Biophysics
New York University Medical Center, New York, New York 10016
[2]Bates College, Lewiston, Maine 04240

The opercular epithelium of <u>Fundulus heteroclitus</u> contains numerous chloride cells identical to the ones of the gills. When mounted in Ussing type chambers th cells transport chloride ions from the basolateral to the sea water side (Zadunaisky, 1984, Fish Physiol. Vol X b). The SCC is an indicator of this Cl^- secretion.

We have utilized the fact that the short circuit current follows closely the changes in intracullular pH to examine the presence of a Na/H exchanger in these cells. In the frog skin epithelium, intracellular pH measurements have indicated that the short circuit current is an excellent parameter for studies of cell pH (Harvey, B.J., et al, J.Gen. Physiol., Vol.<u>92</u> pp 767-791, 1988). The short circuit current is decreased when the intracellular pH becomes acid and it increases as the pH becomes alkaline.

Opercular epithelia from sea water adapted <u>Fundulus heteroclitus</u> were mounted in Ussing chambers and bathed with teleost Ringer solution (see Zadunaisky et. al. MDIBL Bulletin, Vol. 30, 1991, pp 58-59), in which the bicarbonate was substituted with HEPES buffer. The medium was adjusted to pH 7.4.

The tests for the presence of Na^+/H^+ exchanger consisted of acidification with 15mM NH_4Cl added to the basolateral side. There is a transient increase in the SCC as the cell briefly undergoes alkalinization. The SCC then decreases to a minimum during the acidification, at which time the solution in the basolateral side is replaced with fresh Teleost ringer without NH_4Cl. This results in restoration of the normal intracellular pH. The current returns to its initial resting value as the intracellular pH becomes more alkaline. This method, introduced by Boron (boron W.F., DeWeer, F&P, J. Gen. Physiol. Vol. <u>67</u> pp 91-112, 1976) produces a sequence of intracellular pH displacements that activate the Na/H exchanger.

The addition of NH_4Cl to the apical side did not produce changes in the short circuit current, but when added to the basolateral side, a rapid drop in the current was observed. Therefore the additions of NH_4Cl and other drugs were performed only on the basolateral side of the preparation.

Amiloride, at concentrations of 10^{-2} and 10^{-3}M stopped or reduced the rate of recovery from the acid load. At 10^{-2} the remaining current was 20%, of controls in 7 experiments. These results were interpreted as a good indication that acidification had elicited the activation of the Na/H exchanger with subsequent alkalinization. This event was inhibited by amiloride at high concentrations.

The rate of recovery from the NH_4Cl acid load was accelerated by several factors that were tested on the basis of their known effect on the Na/H exchanger in other organs or tissues (Grinstein, S. (Ed) Na/H Exchange, CRC Press, 1988).

Epidermal growth factor, at 10^{-5}M produced an acceleration of the recovery of 155% in 7 experiments. Phorbolester (PMA) at a concentration of 10^{-5}M produced a stiumulation of 165% in 7 experiments and IBMX at 10^{-4}M produced a stimulation of 216% in 8 experiments.

The sequence of events during the pH transients, the inhibition with amiloride at high concentrations, and the stimulation with EGF, PMA and IBMX permit us to conclude that a Na/H exchanger is present in the basolateral membranes of the chloride cells, which is activated by internal protons.

The rapid changes and transients in chloride secretion when small changes in pH are occuring is an indication that intracellular pH must be part of the control and signaling that these cells undergo in order to transport more or less salt. The transport, in turn maintains the homeostasis of the ion concentration, especially NaCl in the plasma of the fish. In fact, in another report in this issue of the Bulletin we find that changes in osmolarity activate the Na/H exchanger of the chloride cells.

Acknowledgements: This work was supported by NIH grant EYO1340 to JAZ.

Abbreviations used: SCC - Short Circuit Current; EGF - Epidermal Growth Factor; IBMX - 3-Isobutyl-1-Methyl-Xanthine.

IDENTIFICATION OF A BASOLATERAL Na/H ANTIPORTER IN URINARY BLADDER OF FLOUNDER, PSEUDOPLEURONECTES AMERICANUS

Marc A. Post and David C. Dawson
Department of Physiology, University of Michigan
Ann Arbor, MI 48109

Na/H antiporters have been identified in the apical and basolateral membrane of epithelial cells. The apical isoform is one component of an electrically silent active Na^+ absorption pathway, but the function of the basolateral isoform of the exchanger is less well understood. We recently showed that the basolateral membrane of turtle colon, an electrogenic Na^+ absorptive epithelium, expresses a high level of amiloride-sensitive Na/H exchange activity (Post & Dawson, FASEB J. 4:A549, 1990). The purpose of these experiments was to determine if Na/H exchange activity could be identified in the basolateral membrane of the flounder urinary bladder, an electrogenic K^+ secretory epithelium. We assayed for the presence of exchange activity by measuring ouabain-insensitive, amiloride-inhibitable transmural fluxes of $^{22}Na^+$ across sheets of bladder which had been apically permeabilized with the pore-forming antibiotic amphotericin-B, and by imposing large transmural gradients of Na^+ or H^+ to induce counterflow or "transacceleration" of the $^{22}Na^+$ flow.

Flounder urinary bladder was initially mounted in Ussing chambers ($0.287 cm^2$) as previously described (Wilkinson & Dawson, Bull. M.D.I.B.L. 29:108-109, 1990). The Ringer's contained (in mM): 140 Na^+, 147.5 Cl^-, 2.5 K^+, 1.5 Ca^{2+}, 1.0 Mg^{2+}, 15 HEPES, and 10 glucose, at a pH of 7.5. 100 µM verapamil was present in the serosal bath to minimize smooth muscle activity and 100 µM ouabain was present to inhibit basolateral Na/K-ATPase activity. Unidirectional $^{22}Na^+$ fluxes were determined in similar solutions except that chloride was replaced by gluconate, and in some cases K^+ replaced all but 2 mM of the Na^+. All Ringer's solutions used for flux determinations were initially adjusted to a pH of 6.5 to ensure maximal activation of the antiporter. Transmural fluxes of $^{22}Na^+$ were measured according to the "sample and replace" paradigm of Dawson (J. Membr. Biol. 37:213-233, 1977), except that 5 µCi $^{22}Na^+$ was used, the interval between samples was shortened to 10 minutes, and the sample volume (1 ml) was 25% of the bath volume (4 ml).

Table 1 presents evidence for basolateral Na/Na exchange in flounder urinary bladder. In the presence of serosal ouabain and a 140 to 2 mM mucosal (M) to serosal (S) Na^+ gradient the unidirectional rate coefficient for S to M $^{22}Na^+$ flow (λ^*_{SM}) was greater than the unidirectional rate coefficient for M to S $^{22}Na^+$ flow (λ^*_{MS}). This result may reflect the presence of apical Na/Na exchange, mediated by the NaCl cotransporter (Stokes, J. Clin. Inv. 74:7-16, 1984), in series with basolateral Na/Na exchange, mediated by a Na/H exchanger. Permeabilizing the apical membrane (10 µM amphotericin B) revealed a large transacceleration of λ^*_{SM} by the M to S Na^+ gradient. The addition of amiloride (0.5 mM) to the serosal bath reduced both λ^*_{SM} and λ^*_{MS} to near zero. The magnitude of the Na/Na exchange activity can be best appreciated by noting that the amiloride-sensitive portion of λ^*_{SM} was 50 times greater than the amiloride-sensitive portion of λ^*_{MS} (i.e $\Delta SM/\Delta MS=50$). These results are consistent with the notion that a high level of amiloride-sensitive Na/Na exchange activity is present in the basolateral membrane of flounder urinary bladder. Because these are well known characteristics of the Na/H exchanger (Aronson, Ann. Rev. Physiol. 47:545-60, 1985), we assayed for Na/H exchange by using a proton gradient to drive $^{22}Na^+$ counterflow.

Figure 1 depicts λ^*_{SM} first in the absence and then in the presence of an M to S proton gradient. Both baths contained K gluconate Ringer's solution at pH 6.5; the serosal bath also contained ouabain (100 µM) and verapamil (100 µM). Initially the opposing unidirectional rate coefficients were low and symmetric (λ^*_{MS} not shown), as expected for ouabain treated tissue in the absence of transmural ion gradients.

Permeabilization of the apical membrane by amphotericin (10 μM) in the absence of a Na$^+$ or H$^+$ gradient revealed that a basolateral route for ^{22}Na$^+$ flow was present (indicated by the increase in λ_{SM}^*. Imposition of an M to S H$^+$ gradient, by alkalinizing the serosal bath to pH 8.5 with KOH, transaccelerated λ_{SM}^*, as expected for obligatory Na/H exchange. The addition of 1 mM amiloride to the serosal bath reduced λ_{SM}^* to below the level measured prior to apical permeabilization, consistent with complete block of the basolateral pathway for Na$^+$ movement.

The results presented in this report support the notion that robust Na/H exchange activity is expressed in the basolateral membrane of winter flounder urinary bladder. As in other tissues, the flounder bladder Na/H exchanger was able to catalyze both Na/Na and Na/H exchange, and was inhibited by amiloride. It will be of interest to determine if the antiporter plays a role in the regulation of transepithelial transport as suggested by Harvey and Ehrenfeld (J. Gen Physiol. 92:793-810, 1988).

This research was supported by grants from NIEHS (ES03828 to David H. Evans), NIH (DK29786 to DCD), and the Cystic Fibrosis Foundation (GF).

Table 1
Unidirectional Rate Coefficients for ^{22}Na$^+$ Flow Across Flounder Urinary Bladder (Mean ± SD)

	λ_{MS}^* (cm/h x 10^3)	λ_{SM}^* (cm/h x 10^3)
Serosal ouabain (100 μM)	1.2 (0.2)	19.9 (3.3)
Mucosal amphotericin-B (10μM)	7.0 (0.8)	185.9 (9.6)
Serosal amiloride (0.5 mM)	3.4 (0.7)	6.9 (2.3)
Δ	2.6	179

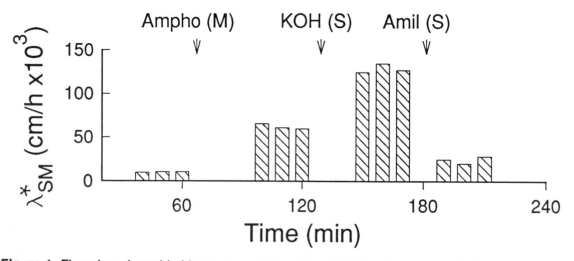

Figure 1 Flounder urinary bladder possesses basolateral Na/H antiporters. KOH addition to the serosal bath imposed an M to S H$^+$ gradient (mucosal pH = 6.5; serosal pH = 8.5) and transaccelerated the rate coefficient for S to M ^{22}Na$^+$ flow (λ_{SM}^*) across amphotericin-B (ampho) permeabilized tissue. Subsequent serosal addition of amiloride (1 mM) reduced λ_{SM}^*.

MOLECULAR BIOLOGY OF THE ELECTROGENIC SODIUM/HYDROGEN ANTIPORTER IN GILLS OF THE GREEN SHORE CRAB CARCINUS MAENAS

David W. Towle[1], Marek Kordylewski[2], Susann W. Bowring[1]
and Alison I. Morrison-Shetlar[3]
[1]Department of Biology, Lake Forest College, Lake Forest, IL 60045
[2]Harvard University, Dunster House A34, Cambridge, MA 02138
[3]Max-Planck-Institut für Systemphysiologie, 4600 Dortmund 1, Germany

The Na^+/H^+ antiporter in crustacean gill epithelia is uniquely electrogenic, apparently exchanging 2 Na^+ for 1 H^+, unlike the electroneutral 1 $Na^+/1$ H^+ antiporter in vertebrate tissues (Shetlar, Alexander and Towle, Bull. MDIBL 27:59, 1988; Shetlar and Towle, Am. J. Physiol. 257:R924, 1989). The 2:1 stoichiometry has been confirmed in other crustacean as well as echinoderm tissues (Ahearn and Clay, Am. J. Physiol. 257:R484, 1989; Ahearn and Franco, J. Exp. Biol. 158:495, 1991). The vertebrate "housekeeping" Na^+/H^+ antiporter has recently been cloned and sequenced from mammalian tissues (Sardet, Franchi and Pouysségur, Cell 56:271, 1989; Tse et al., EMBO J. 10:1957, 1991). Due to the generosity of Dr. Jacques Pouysségur who provided a cDNA fragment containing 80% of the human Na^+/H^+ antiporter sequence, we have been able to investigate whether discernible sequence homology exists between the vertebrate and crustacean antiporter genes.

Genomic DNA and total RNA were isolated from male gonad and gill respectively of the green shore crab Carcinus maenas according to the methods of Miller, Dykes and Polesky (Nucl. Acids Res. 16:1215, 1988) and Chomczynski and Sacchi (Anal. Biochem. 162:156, 1987). Aliquots of crab DNA and RNA were applied to nylon membrane using a slot-blot apparatus (BioRad Laboratories). The membranes were then hybridized overnight with the human Na^+/H^+ antiporter cDNA probe which had been biotinylated using the Flash™ labelling and detection system provided by Stratagene Cloning Systems. Under conditions of moderate stringency, hybrids could be detected between crustacean and human nucleotide sequences, indicating partial sequence homology of the antiporter genes.

Several oligonucleotide primers were synthesized based on putative transmembrane and intracellular regions of the human Na^+/H^+ antiporter. Using crab genomic DNA as template, primers representing transmembrane sequences supported polymerase chain reaction amplification of the DNA, producing discrete bands discernible upon agarose gel electrophoresis. However, primers representing intracellular regions did not support amplification, suggesting that only transmembrane sequences of the Na^+/H^+ antiporter may exhibit partial homology between human and crab.

Supported by National Science Foundation DCB-8996137, DCB-9024293, and a Research Experiences for Undergraduates Supplement.

DISTRIBUTION OF THE Na-K-Cl COTRANSPORTER
IN THE SPINY DOGFISH, SQUALUS ACANTHIAS

Biff Forbush, John Payne, Jian-Chao Xu, Chris Lytle, Ed Benz,
Jocelyn Forbush, Tracey Tong Zhu, and Grace Jones
Department of Cellular and Molecular Physiology
Yale University School of Medicine, 333 Cedar St., New Haven, CT 06510

To investigate the distribution of the Na-K-Cl cotransport protein in various tissues of the dogfish shark we have used both monoclonal antibodies (C. Lytle, J-C. Xu, T.T. Zhu, M. Haas, B. Forbush H, J. Gen. Physiol. 96:44a, 1990; and Lytle et al. in preparation), and cDNA probes (Xu et al., in preparation) specific for the Na-K-Cl cotransporter. In rectal gland the cotransport protein has been identified as a ~195 kDa protein by photolabelling with a derivative of benzmetanide (B. Forbush III and M. Haas, Biophys. J., 53:222a, 1988). Much of the apparent mass is carbohydrate, since when treated with N-glycanase the deglycosylated protein molecular weight is found to be ~137 kDa. From the number of $[^3H]$benzmetanide binding sites it is estimated that the Na-K-Cl cotransporter comprises approximately 2% of the total membrane protein.

We have utilized four monoclonal antibodies -- each of these reacts with a distinct epitope on the 195 kDa protein, each is capable of specific immunoprecipitation of $[^3H]$benzmetanide reversibly bound to the protein, and each is effective in both Western blotting and immunohistochemical procedures. There are individual characteristics: J4 shows a strong preference for native protein and reacts relatively weakly on Western blots and only with the intact 195k Da protein, whereas J3 and J7 show a preference for denatured protein and react strongly on Western blots with the intact protein as well as with a number of large and small proteolytic fragments of the protein. J25 reacts with a large protease-resistant portion of the protein, and its weak reactivity with N-glycanase-treated protein suggests that its epitope may be part of the oligosaccharide moiety.

To estimate the relative amounts of Na-K-Cl cotransporter in various tissues, we examined the binding of ^{35}S-labelled antibodies to isolated membranes. Membranes were prepared from a number of shark tissues by homogenization and differential centrifugation in the presence of protease inhibitors (cf. B. Forbush III, M. Haas, and C. Lytle, Am. J. Physiol., in press). In some experiments the membranes were treated briefly with BSA-buffered SDS (0.7 mg/ml SDS, 1% BSA) to ensure that any membrane vesicles were broken. Membrane suspensions were incubated with ^{35}S-labelled antibodies (2 µg/ml) for 1 hour at 20°C in PBS-1%BSA -- bound antibody was determined by dilution and filtration of the samples followed by scintillation counting.

In a typical experiment, the relative amounts of ^{35}S-J3 antibody binding to membranes from various tissues were: rectal gland=1.0, ciliary fold (eye) 0.11, retinal pigment epithelium 0.078, gall bladder 0.068, brain 0.053, lens epithelium 0.049, testes 0.023, intestine 0.022, kidney 0.014, liver 0.014, gill 0.013, spleen 0.012, stomach (epithelial scraping) 0.009, skeletal muscle 0.002, pancreas 0.0008. Because of the wide differences in dissection, membrane purification, and cell-type homogeneity from tissue to tissue these values can only be taken to give a rough impression of the amount of cotransporter in various tissues. It can be noted that as judged by antibody binding the rectal gland contains 10- to 100-fold more transport protein as other tissues. This is consistent with the previous observation that there are at least 50 times more [^{3}H]benzmetanide binding sites in rectal gland membranes compared to membranes from most other tissues (eg. dog kidney membranes; Forbush et al., ibid.). Comparing antibody ^{35}S-J4 to antibody ^{35}S-J3, we found that J4 exhibits greater tissue specificity -- J4 was 2-5 times less immunoreactive compared to J3 in all tissues except brain and rectal gland.

Western blot analysis of antibody reactivity was carried out using standard procedures. Approximately 50 μg of SDS-solubilized membrane protein was loaded per gel lane (except rectal gland, 0.5 μg). With the luminol-peroxidase system (Amersham ECL) as little as ~1 ng of the 195 kDa protein was detected on nitrocellulose blots using the J3 antibody. Confirming the result obtained with ^{35}S-labelled antibodies, there appeared to be approximately 20-100 times as much reactivity in rectal gland membranes as in other tissues. In different tissues, we found that proteins of three different molecular weights were labelled.

A 195k Da protein was identified in rectal gland membranes with each of the antibodies. This band was not detected in any of the other tissues. However in membranes from brain and the eye, a 160 kDaprotein was detected by all of the antibodies, which on deglycosylation with N-glycanase was identical in size to the deglycosylated rectal gland protein. The brain protein also yielded some immunoreactive partial tryptic fragments of the same size as those from rectal gland.

In membranes from other tissues (ie. gill, liver, intestine, kidney, stomach, gall bladder, spleen) a >250 kDa protein was the only protein detected. This high molecular weight species was seen with all antibodies except J4, confirming the above-noted tissue specificity of J4. As determined by Western blot analysis, the protein was completely resistant to N-glycanase and trypsin at ten times the concentration needed to digest the 195 kDa and 160 kDa proteins.

The results of the Western blot analysis suggest that there may be different isoforms of the Na-K-Cl cotransport protein, but the results would also be consistent with the hypothesis that the same polypeptide was differentially glycosylated in different tissues. Accordingly we performed Northern blot analysis to see if the mRNA encoding the polypeptide was the same from tissue to tissue. We used two non-overlapping DNA probes consisting of a 1.6 Kb fragment at the 5′ end of the coding region and a 1.9Kb fragment from the 3′ end of the ~3900 Kb coding region (Xu et al., in preparation) and low stringency conditions (2 washes, 37℃) to analyze poly(A)-selected RNA from various tissues. Each of the two probes recognized the following bands: rectal gland 7.3 Kb; brain 7.0 Kb; gill, kidney, liver, spleen, testes, 13 Kb. Differences were noted in the relative strength of hybridization between the two probes, the probe for the 5′ region reacting relatively weakly with the 13 Kb species.

These results illustrate a broad tissue distribution of the Na-K-Cl cotransporter, as indicated in numerous previous physiological studies. It is clear from these experiments that there are different forms of the transport protein in different tissues, and it is strongly suggested that the protein differs both in the polypeptide backbone and in the attached oligosaccharide.

Supported by NIH-DK17433, a CT-AHA grant-in-aid and a CT-AHA fellowship (to C. Lytle).

REGULATION OF Na-K-Cl COTRANSPORT IN THE Cl-SECRETING CELLS OF THE SHARK (SQUALUS ACANTHIAS) RECTAL GLAND

Christian Lytle, Neils Ringstad, Christina Forbush, and Biff Forbush
Department of Cellular and Molecular Physiology
Yale University School of Medicine
New Haven, CT 06510

The rectal gland secretes a salty fluid profusely when exposed to agents which enrich intracellular [cAMP] (forskolin, vasoactive intestinal peptide, adenosine). Models of epithelial secretion customarily envision chloride exit as the primary determinant of secretory output, a belief bolstered by the recent discovery of apical chloride channels activated by cAMP via protein kinase A (Greger, Schlatter, and Gogelein, 1985, Pflügers Arch. 403: 446-448). However, since transmembrane ion gradients change little during rectal gland secretion (Greger, Schlatter, Wong, and Forrest, 1984, Pflügers Arch. 402: 376-384), the cell must replenish chloride ions as soon as they are lost without an increase in the force driving inward Na-K-Cl cotransport. The compensatory chloride uptake must therefore involve activation or recruitment of otherwise dormant Na-K-Cl cotransport units.

Our test of this prediction employed secretory tubules enzymatically liberated from thin rectal gland slices, and two probes: (1) [^{3}H]benzmetanide, a potent inhibitor of Na-K-Cl cotransport which binds avidly only to active cotransporters, and (2) a panel of monoclonal antibodies which selectively recognize the 200 kDa cotransport protein. Our studies on intact perfused glands (Forbush, Haas, and Lytle, 1992, Am. J. Physiol. in press) and isolated tubules (Lytle and Forbush, 1992, Am. J. Physiol. in press) established that secretagogues evoke a dramatic (13-fold) increase in [^{3}H]benzmetanide binding site density. Thus, hormonal modulation of rectal gland secretion must involve a coordinated activation of chloride entry (via basolateral Na-K-Cl cotransporters) and exit (via apical Cl channels).

Binding studies also revealed that the cotransporter is activated by osmotically-induced increases or decreases in cell volume (Lytle and Forbush, 1992, ibid.). In fact, a 45% reduction in cell water content proved to be as effective as secretagogues in promoting [^{3}H]benzmetanide binding, and a 40% increase in cell water was half as effective.

Our recent studies suggested that the functional residency and/or turnover rate of the cotransport protein is controlled by direct phosphorylation (Lytle and Forbush, 1990, J. Cell Biol. 111:312a). Using monoclonal antibodies that selectively immunoprecipitate the 195 kDa cotransport protein (Lytle, Xu, Zhu, Haas, and Forbush, 1990, J. Gen. Physiol. 96:44a), we found that secretagogues or osmotic perturbations produce parallel changes in cotransporter activation state and phosphorylation state (i.e. tubule [^{3}H]benzmetanide binding site density and cotransporter ^{32}P content co-vary).

How chloride entry keeps pace with chloride exit in the face of 30-fold changes in transcellular chloride flow is poorly understood. The sensitivity of the cotransport process to both cAMP and cell volume raised the possibility that hormonal activation of chloride entry represents a corrective response to cell shrinkage following primary activation of Cl channels by cAMP. Our results, however, did not support the "volume-coupling" model. First, to produce the amount of cell shrinkage required to raise [^{3}H]benzmetanide binding to the level evoked by secretagogues (via isotonic KCl loss), the rectal gland cell would have to jettison at least 3 times as much chloride as it possesses. Second, gravimetric measurements of cell water indicated that the rectal gland cell does not experience large sustained shifts in cell volume during secretion as predicted by the volume-coupling model.

An attractive alternative is that cytoplasmic chloride both participates in and modulates the cotransport process. Studies with ion selective microelectrodes do in fact suggest that [Cl]$_i$ declines slightly upon secretion (Greger et al., 1984, ibid.). A sensitivity to [Cl]$_i$ itself would account for the apparent activation of cotransporters by maneuvers known to reduce [Cl]$_i$, such as exposure of the rectal gland cell to furosemide (Greger et al., 1984, ibid.), to low $[Cl]_o$ (Forbush et al., 1992, ibid.), or to low osmolality (Lytle and Forbush, 1992, ibid.). It would also account for the observation that barium, which prevents conductive K exit (and therefore Cl exit) during secretion, blocks cotransporter activation by secretagogues (Forbush et al., 1992, ibid.). Our hypothesis that $[Cl]_i$ modulates cotransport is supported by four additional observations:

First, we found that pre-incubation of rectal gland tubules in media lacking Na or Cl (NMDG or gluconate substitutions, respectively), maneuvers which reverse the normal direction of Na-K-Cl cotransport and lower $[Cl]_i$, increased [³H]benzmetanide binding site density. This response clearly differs from the stimulatory effect of substrate ions on loop diuretic binding since all [³H]benzmetanide binding assays were conducted in normal shark Ringers over a brief interval immediately folowing pre-incubation. Moreover, this effect of ion omission developed gradually over a 10 min interval, was rapidly reversible, and elevated binding to a level about half that produced by secretagogues alone.

Second, the activation of cotransporters by secretagogues was highly sensitive to external potassium — for example, raising $[K]_o$ from 5 to 80 mM abolished the effect of forskolin on [³H]benzmetanide binding to isolated tubules. The inhibition by K_o was half-maximal at 40-50 mM, reversible, and unrelated to its role as a cotransported substrate (K only stimulated binding to isolated membranes). Besides blocking cAMP-induced [³H]benzmetanide binding, high $[K]_o$ blocked cAMP-induced cotransporter phosphorylation. High $[K]_o$ is also known to depolarize the membrane potential (Greger et al., 1984, ibid.), to elevate intracellular Cl, and to enlarge the rectal gland cell (Kleinzeller et al., 1985, J. Comp. Physiol. 155: 145). However, the effect of high $[K]_o$ on cotransporter activition cannot be due to cell swelling since (a) swelling itself activates the cotransporter, and (b) K_o causes the cell to swell gradually but blocks cotransporter activation immediately. An established link between $[K]_o$ and cotransporter regulation is $[Cl]_i$ — since K_o depolarizes the membrane potential, it should prevent or retard chloride exit through apical cAMP-induced Cl channels.

Third, cotransporter activation by cell shrinkage, like that by secretagogues, was blocked by raising $[K]_o$, but at 5-fold lower concentrations (IC_{50} ~10 mM). Importantly, this inhibition by K_o required both Na and Cl externally. An explanation for this phenomenon emerged from estimates of intracellular ion concentrations before and after cell shrinkage — when the medium was rendered hypertonic by the addition of 580 mM sucrose, the cells lost about half their water by osmosis, cytoplasmic ions were concentrated, and the net driving force for Na-K-Cl cotransport shifted from one overwhelmingly favoring salt entry to one favoring salt loss. We conclude that high K_o acts by fortifying the driving force for inward cotransport — this, in turn, maintains $[Cl]_i$ high enough to suppress activation of auxiliary cotransporters.

Fourth, when nominally-active cotransporters in quiescent rectal gland tubules were inhibited with the loop diuretic bumetanide, a maneuver which reduces $[Cl]_i$ (Greger et al., 1984, ibid.), the phosphorylation state of the cotransport protein increased 3-fold within 10 min. Tubules exposed to hypertonicity, secretagogue, or low $[Cl]_o$ for this duration exhibited a 4.3-fold, 8.8-fold, or 5.2-fold increase in cotransporter phosphorylation (n = 3).

In summary, our observations suggest that: (1) cytoplasmic chloride depletion activates the Na-K-Cl cotransport protein by promoting its phosphorylation; (2) hormonal activation of cotransport represents a corrective response to a reduction in $[Cl]_i$ following primary activation of apical Cl channels by cAMP; (3) activation of the cotransporter by cell swelling is a response to intracellular chloride dilution; (4) activation of the cotransporter by cell shrinkage is a response to a reduction in $[Cl]_i$ via outward Na-K-Cl cotransport; (5) $[Cl]_i$ may be the signal that coordinates chloride entry and exit during secretion.

Supported by the American Heart Association, a grant (DK17433) and postdoctoral fellowship (DK07259) from the NIH, and a Blum-Halsey Scholar Award (to C. Lytle)

SKATE (Raja erinacea) BASOLATERAL LIVER PLASMA MEMBRANES EXHIBIT AMINO ACID TRANSPORT SYSTEM A-LIKE ACTIVITY

S. A. Ploch and D. J. Smith
Department of Medicine and Environmental Toxicology Center,
University of Wisconsin, Madison, WI 53706

The liver is the predominant site for the metabolism of amino acids. In mammals the transport of alanine and possibly glutamine across the hepatocyte plasma membrane appears to be rate limiting in their metabolism. Although skates (Raja erinacea) like many other fish consume a high protein diet, little is known regarding neutral α-amino acid transport by the liver of this marine elasmobranch. A transport system for β-alanine has been identified in the skate liver (Shuttleworth, J. Exp. Zool. 231:39, 1984). Using isolated skate hepatocytes, Ballatori has shown that alanine is transported by at least two sodium-dependent transport systems (Am. J. Physiol. 254:R801, 1988). However, it is sometimes difficult to separate plasma membrane transport processes from subsequent metabolic steps using isolated cell preparations. Also in a polarized cell like the hepatocyte, certain transport systems are located exclusively at the basolateral or apical pole of the cell. Therefore we have begun to investigate α-amino acid transport by isolated skate basolateral liver plasma membrane vesicles.

As previously described, skate basolateral liver plasma membranes were isolated on self-generating Percoll gradients (Smith, J. Exp. Zool. 258:189, 1991). The amino acid transport activity of these basolateral liver plasma membrane vesicles was assayed using a rapid filtration technique. The aminotransferase inhibitor aminooxyacetic acid was included in the transport assay at a concentration of 2 mM.

Uptake of alanine and its unnatural analogue, α-methylaminoisobutyric acid (MeAIB) was stimulated and demonstrated an over-shoot in the presence of an out-to-in 100 mM transmembrane sodium gradient (Figure 1); alanine (50 μM) uptake was linear for at least 15 sec and MeAIB (50 μM) uptake was linear for at least 30 sec. Sodium-dependent MeAIB uptake was completely inhibited by 20 mM alanine. In contrast, 20 mM MeAIB produced only partial (67%) inhibition of sodium-dependent alanine uptake. These results are similar to those of Ballatori in that they indicate that sodium-dependent alanine transport by the skate liver occurs via at least two transport systems, one of which it shares with MeAIB. Preliminary characterization of the component of sodium-dependent alanine transport which is not inhibited by MeAIB indicates that this portion of alanine uptake is significantly inhibited by 20 mM cysteine, serine, glutamine, threonine and sarcosine, but only minimally reduced by 20 mM leucine, β-alanine and taurine.

In mammals MeAIB is a specific substrate for amino acid transport system A. Therefore MeAIB uptake by skate basolateral liver plasma membrane vesicles was examined further. Preliminary kinetic analysis of sodium-dependent MeAIB uptake suggests the presence of a high affinity system with an approximate K_m of 93 μM. The MeAIB transport activity of these membrane vesicles was pH sensitive. The initial rate of MeAIB transport was maximal (39.2$\pm$4.7 pmol/mg prot/30 sec) at pH 8.0 and decreased progressively with either an increase or decrease in pH. Physiological levels of urea in the skate are relatively high (398 mM); this level of urea has the potential to denature proteins (Yancey, Science 217:1214,

1982). It has been proposed that the high level of trimethylamine-N-oxide (TMAO) found in marine elasmobranchs may protect proteins from the de-naturing effects of urea. Therefore we investigated the effect of urea (398 mM), TMAO (120 mM) or the two together on the initial rate of MeAIB uptake. Neither agent alone nor the two together altered the initial rate of MeAIB uptake.

In summary, the skate basolateral liver plasma membrane contains two transport systems for sodium-dependent alanine uptake. One of these transport systems shares many characteristics with the mammalian amino acid transport system A, e.g. it transports MeAIB and is pH sensitive. The majority of sodium-dependent alanine transport by these membrane vesicles occurred via this system A-like transporter. The system A-like transporter is resistant to the perturbing effects of urea; the resistance of the transporter to the effects of this concentration of urea is somewhat unusual among proteins, even those from urea rich species (Yancey, Science 217:1214, 1982).

FIGURE:

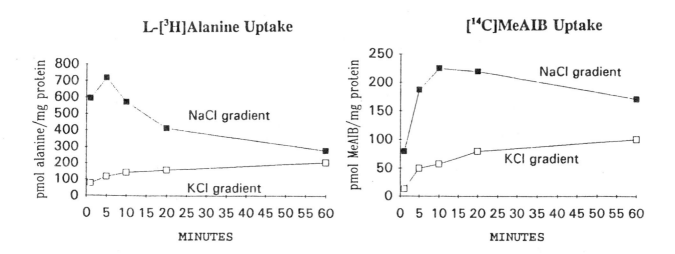

Supported in part by NIH grant DK 38883.

A ROLE FOR PROTEIN KINASE C IN THE CONTROL OF K SECRETION AND Na ABSORPTION BY THE URINARY BLADDER OF THE WINTER FLOUNDER, PSEUDOPLEURONECTES AMERICANUS

Marc A. Post, Greg Feero, and David C. Dawson
Department of Physiology
University of Michigan
Ann Arbor, MI 48109

The urinary bladder of the winter flounder absorbs NaCl and secretes K^+. The absorptive process is electrically silent presumably due to the presence in the apical membrane of a thiazide sensitive NaCl cotransporter (Stokes, J. Clin. Inv. 74:7-16, 1984). In contrast, K^+ secretion is associated with a serosa to mucosa current which is a measure of the rate of conductive K^+ exit across the apical membrane via barium-sensitive K^+ channels (Dawson & Frizzell, Pflügers Arch 414:393-400, 1989). Although the apical transport steps for the absorptive and secretory processes have been characterized, there are, as yet, no known hormonal regulators of salt transport in this tissue, and nothing is known about intracellular events which might regulate ion flows at the apical or basolateral membrane. The purpose of these experiments was to investigate the possibility that the ubiquitous enzyme, protein kinase C, might have some role in the regulation of salt transport by the bladder. The results provide evidence for inhibitory control of both NaCl absorption and K^+ secretion.

Urinary bladder from winter flounder was mounted as a flat sheet as previously described in either Ussing chambers (0.287 cm^2) for $^{22}Na^+$ flux studies or in perfusion chambers for electrophysiological studies in the absence of radioisotope (Wilkinson & Dawson, Bull. M.D.I.B.L. 29:108-109, 1990). All fluxes were measured under short circuit conditions (serosal bath as reference) using the sample and replace paradigm of Dawson (J. Membr. Biol. 37:213-233, 1977), modified as previously described (Post & Dawson, Bull. M.D.I.B.L., this issue). Solutions for flux measurements consisted of (in mM): 140 Na^+, 147.5 Cl^-, 2.5 K^+, 1.5 Ca^{2+}, 1.0 Mg^{2+}, 15 HEPES, and 10 glucose, at a pH of 7.5. 100 µM verapamil was present in the serosal bath to minimize smooth muscle activity. Solutions for perfusion experiments were similar except that, where indicated, 10 mM K Gluconate was added to the mucosal perfusate to effect changes in mucosal K^+ concentration.

Figure 1 presents evidence that phorbol esters inhibit both K^+ secretion and Na^+ absorption. I_{SC}, a direct measure of K^+ secretion (Dawson & Frizzell, Pflügers Arch 414:393-400, 1989), and the mucosal to serosal $^{22}Na^+$ rate coefficient ($\lambda_{MS}^{\bullet}$), a measure of Na^+ absorption (Stokes, J. Clin. Inv. 74:7-16, 1984), are plotted versus time for a bladder secreting K^+. The polarity of the current was consistent with positive charge movement from the serosal (S) to the mucosal (M) bath. The rate coefficient for $^{22}Na^+$ flow was indicative of a transmural pathway for M to S Na^+ movement. Addition of 10 nM 4-β-phorbol 12,13 myristate acetate (PMA) to the mucosal bath decreased both I_{SC} and $\lambda_{MS}^{\bullet}$. The relatively minor effect of the subsequent mucosal addition of 100 µM hydrochlorothiazide (HCT), a known inhibitor of

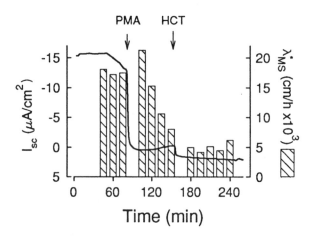

Figure 1 10 nM 4-β-phorbol 12,13 myristate acetate (PMA) inhibited most of the hydrochlorothiazide (HCT) sensitive Na^+ absorption and K^+ secretion in flounder urinary bladder.

the NaCl cotransporter (Stokes, J. Clin. Inv. 74:7-16, 1984), is consistent with the notion that the phorbol ester eliminated most of the NaCl cotransport activity as well as K^+ secretion.

Inhibition of electroneutral NaCl cotransport by HCT or Hg^{2+} inhibits K^+ secretion , presumably by decreasing the driving force for K^+ exit across the apical membrane (Wilkinson & Dawson, Bull. M.D.I.B.L. 29:108-109, 1990). Thus PMA could inhibit K+ secretion either indirectly by inhibiting the NaCl cotransporter or by a direct effect on the apical K^+ conductance (or both). Figure 2 shows I_{sc} for a urinary bladder mounted in a perfusion chamber. Apical K^+ conductance was assessed by determining the effect of a 10mM step increase of the mucosal K^+ concentration (indicated in figure 2 as a solid bar along the time axis). As shown in the figure, 0.5 µM Hg^{2+} reduced I_{sc} but, as indicated by the response to the step increase in mucosal K^+ concentration, the apical K^+ conductance was, if anything, increased as has been previously reported. In contrast, after inhibition of I_{sc} by 10 nM PMA apical K^+ conductance was reduced or absent. Exposure to up to 1 µM 4-α-phorbol 12,13 didecanoate, an inactive phorbol ester, was without effect on either I_{sc} or G_T (not shown). These results support the notion that PMA decreased K^+ secretion via inhibition of the apical K^+ conductance.

Figure 3 shows that 100 µM 1,2 dioctanoyl-sn-glycerol (C:8) (DAG), a diacylglycerol analogue, also inhibited both Na^+ transport and K^+ secretion. The subsequent addition of 100 µM HCT to the mucosal bath revealed that the DAG exerted a more striking effect on K^+ secretion than on Na^+ absorption.

The profound inhibitory action of 10 nM PMA and the weaker effects of 100 µM DAG contrasts sharply with the ineffectiveness of 1 µM 4-α-phorbol 12,13 didecanoate at inhibiting transmural ion movement. This pharmacological profile is consistent with a mechanism of action involving protein kinase C (Evans et al., Bioch. Soc. Trans. 19:397-402, 1991). The data presented in this report are consistent with the notion that both apical K^+ channels and thiazide sensitive NaCl cotransporters are under the inhibitory control of protein kinase C.

This research was supported by grants from NIEHS (ES03828 to David H. Evans), NIH (DK29786 to DCD), and the Cystic Fibrosis Foundation (GF).

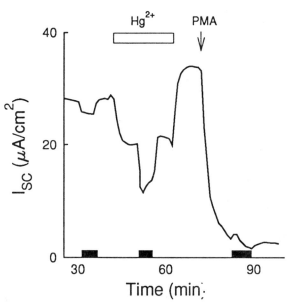

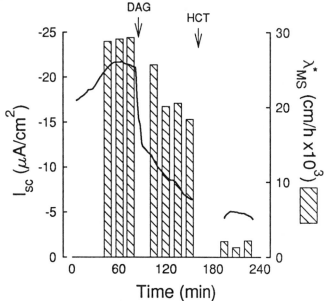

Figure 2 Phorbol esters (10 nM) reduced the apical K^+ conductance of flounder urinary bladder.

Figure 3 100 µM 1,2 dioctanoyl-sn-glycerol (C:8) inhibited both K^+ secretion and Na^+ absorption by flounder urinary bladder.

INORGANIC MERCURY INHIBITS NaCl COTRANSPORT IN FLOUNDER (PSEUDOPLEURONECTES AMERICANUS) URINARY BLADDER

Marc A. Post, Greg Feero, and David C. Dawson
Department of Physiology, University of Michigan
Ann Arbor, MI 48109

In a previous report (Wilkinson and Dawson, Bull. M.D.I.B.L. 29:108-109; 1990) we presented evidence that inorganic mercury inhibited K^+ secretion by the flounder urinary bladder via an indirect mechanism of action: blockade of apical thiazide-sensitive NaCl cotransport. Here we present the results of transmural flux measurements designed to test directly for an effect of divalent mercury on NaCl absorption. The results suggest that Hg^{2+} may be a relatively selective and reversible inhibitor of the apical NaCl cotransporter.

Sheets of flounder urinary bladder were mounted in Ussing chambers as previously described (Wilkinson & Dawson, Bull. M.D.I.B.L. 29:108-109, 1990), and bathed by a Ringer's solution which contained (in mM): 140 Na^+, 147.5 Cl^-, 2.5 K^+, 1.5 Ca^{2+}, 1.0 Mg^{2+}, 15 HEPES, and 10 glucose, at a pH of 7.5. 100 µM serosal verapamil was also present to inhibit smooth muscle activity. The tissue was short circuited (serosal bath as reference) and the current (I_{SC}) and conductance (G_T) were continuously monitored. The I_{SC} in this tissue results directly from K^+ secretion (Dawson & Frizzell, Pflügers Arch 414:393-400, 1989). The mucosal to serosal rate coefficient for $^{22}Na^+$ flow (λ_{MS}^*) was determined as described previously (Dawson & Andrew Bull M.D.I.B.L. 19:46-49, 1979), except that the flux periods were shortened to 15 min. The rate coefficient was calculated as: $\lambda_{MS}^* = J_{MS}^*/C_M^*$, where J_{MS}^* is the mucosal to serosal flux of tracer and C_M^* is the concentration of tracer in the "hotside" bath (in cpm/ml). λ_{SM}^* was determined in a similar fashion. λ_{MS}^* is a direct measure of Na^+ absorption via the hydrochlorothiazide (HCT) sensitive NaCl cotransporter (Stokes, J. Clin. Invest. 74:7-16, 1984)

Figure 1 is a typical experiment (n=8) showing the effect of Hg^{2+} on Na^+ absorption and K^+ secretion in a single tissue. Each bar represents a 15 minute flux period. The addition of 1.5 µM $HgCl_2$ to the mucosal bath produced about a 50% inhibition of Na^+ absorption and a profound inhibition of I_{SC}. This inhibition was at least partially reversed by the mucosal addition of 1 mM dithiothreitol (DTT) to chelate the Hg^{2+}. Subsequent mucosal addition of 100 µM HCT, a specific inhibitor of the NaCl cotransporter (Stokes, J. Clin. Invest. 74:7-16, 1984) markedly attenuated the absorptive Na^+ flux and the I_{SC}. The S to M fluxes of $^{22}Na^+$ (not shown) were unaffected by these maneuvers.

These results are consistent with the hypothesis that mercury reversibly inhibits apical NaCl cotransport in flounder urinary bladder. In separate experiments we characterized the selectivity and reversibility of Hg^{2+} block by monitoring the change in I_{SC} in perfused tissues, as previously described (Wilkinson & Dawson, Bull. M.D.I.B.L. 29:108-109, 1990). Inhibition of I_{SC} by 1 µM Hg^{2+} was completely reversible within 30 min of exposure, but after 90 min of exposure to Hg^{2+} recovery was only about 50% reversible. Other divalent cations such as Ni^{2+} (5 µM), Cu^{2+} (3 µM), Co^{2+} (4 µM), or Cd^{2+} (2 µM) were without effect. In addition, the organic mercurial p-chloromercuribenzenesulfonic acid (PCMBS) had no effect at a concentration of 50 µM, although a slight inhibition of I_{SC} was noted at 150 µM.

These and previous observations can be organized into a speculative model for the action of Hg^{2+} on the flounder urinary bladder. The efficacy of Hg^{2+} and the lack of effect of PCMBS at inhibiting transport suggests that the divalent form of mercury is the active agent. The complete and rapid reversal of the effects of Hg^{2+} either by chelation (fig 1) or by washout in perfusion chambers (not shown) suggests an extracellular site of action. Divalent mercury could act by binding to a site on the HCT sensitive cotransporter (for example, the Na^+ binding site) and thus inhibit the translocation of NaCl. The consequent reduction in Na^+ entry indirectly attenuates apical K^+ exit, perhaps by means of a hyperpolarization of the apical membrane potential as suggested by the results of Duffy and Frizzell (Fed. Proc., 43:444, 1984).

This research was supported by grants from NIEHS (ES03828 to David H. Evans), NIH (DK29786 to DCD), and the Cystic Fibrosis Foundation (GF).

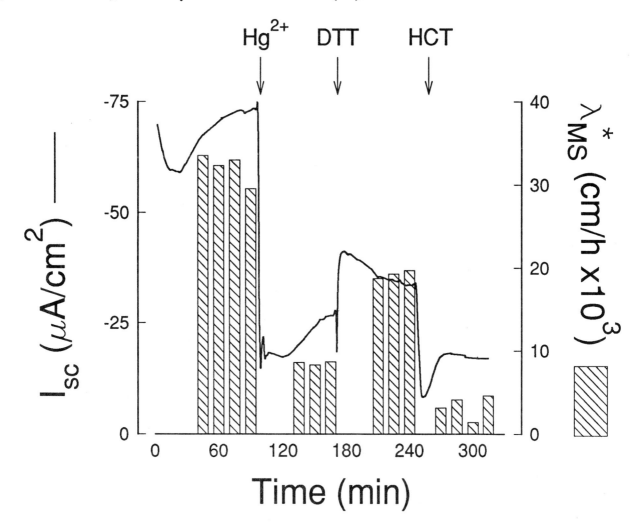

Figure 1 Mucosal addition of 1.5 µM Hg^{2+} inhibited and 1 mM dithiothreitol (DTT) partially restored Na^+ absorption (right hand axis; unidirectional rate coefficient for M to S $^{22}Na^+$ movement, λ_{MS}^*) and K^+ secretion (left hand axis) by the flounder urinary bladder.

TRANSPORT OF URATE INTO HEPATOPANCREATIC BASOLATERAL MEMBRANE VESICLES ISOLATED FROM THE AMERICAN LOBSTER (Homarus americanus)

Anne Nies[1], Eva-Maria Kinne-Saffran[2], Rolf K.H. Kinne[2] and Manfred K. Grieshaber[1]

[1]Heinrich-Heine-Universität Düsseldorf, Universitätsstr. 1,
D-4000 Düsseldorf 1, Germany
[2]Max-Planck-Institut für Systemphysiologie, Rheinlanddamm 201,
D-4600 Dortmund 1, Germany

The hepatopancreas is the major site of urate catabolism in crustaceans. This reaction is an intracellularly located, oxygen-dependent process catalyzed by uricase (Sharma ML & Neveau MC, Comp Biochem Physiol 40B: 863-870, 1969; Noguchi et al., J Biol Chem 254: 5272-5275, 1979). Under hypoxic conditions in vitro urate is released by the hepatopancreas into the incubation medium (Grieshaber, unpublished results). These results suggest the existence of a mechanism in crustacean hepatopancreas which mediates transport of urate across the basolateral membrane. In the present study it was therefore attempted to demonstrate and characterize an urate transport mechanism in the basolateral membrane of lobster hepatopancreas.

Hepatopancreatic basolateral membrane vesicles (BLMV) of the American lobster Homarus americanus were prepared from fresh tissue according to a method described by Ahearn et al. (Am J Physiol 252: R859-R870, 1987) by sorbitol gradient centrifugation. Purity of each membrane preparation was assessed by measuring the activity of marker enzymes such as alkaline phosphatase for apical membranes (Michell et al., Biochem J 116: 207-216, 1970), Na-K-activated ATPase for basolateral membranes (Schwartz et al., J Pharmacol Therapeut 168: 31-41, 1969) and succinate dehydrogenase for mitochondria (Green DE & Ziegler DM, in: Colowick SP, Kaplan NO eds., Methods in Enzymology 6: 416-427, Academic Press, New York). Protein contents of homogenate and vesicle preparation were determined by the method of Lowry et al. (J Biol Chem 193: 265-275, 1951) with bovine serum albumin as the standard.

Transport experiments were carried out as timecourse studies using the Millipore filtration technique (Hopfer et al., J Biol Chem 248: 25-32, 1973). [2-^{14}C]urate (Amersham, Arlington Heights, Il, USA, sp act 49.1 mCi/mmol, 1.81 GBq/mmol) was a generous gift of Dr. Ruth Abramson (Mt. Sinai Medical Center, NY, USA). Throughout this study values are given as means and their standard errors of at least two experiments. Within a given experiment data were sampled in duplicate.

The sorbitol gradient centrifugation according to Ahearn et al. (see above) yielded membranes with enriched Na-K-ATPase activity (8.6 ± 2.6-fold over homogenate activity) but low activity of alkaline phosphatase and succinate dehydrogenase (1.2 ± 0.6-fold and 1.1 ± 0.8-fold, respectively). These results suggest that the basolateral membranes were minimally contaminated by either apical membranes or mitochondria.

Uptake of urate by untreated and preshrunken hepatopancreatic basolateral membrane vesicles was measured in the presence of an initial inwardly directed cation gradient of 100 mM NaCl (figure 1). Uptake increased in curvilinear fashion reaching a plateau after 1 min of incubation at approximately 770 pmol urate/mg protein in the case of untreated and at 280 pmol urate/mg protein in case of preshrunken vesicles indicating transmembrane movement of urate. Uptake of urate into BLMV after equilibration for 90 min was decreased by 62% in the presence of 300 mM sucrose in the extravesicular space.

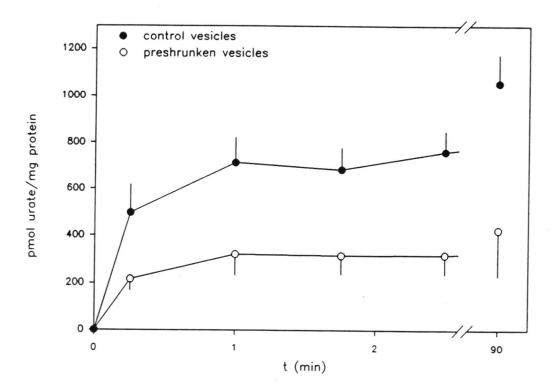

Figure 1. Time course of urate uptake by hepatopancreatic BLMV. Vesicles were loaded with 100 mM mannitol, 10 mM Tris-HEPES, pH 7.6 and were untreated or pretreated for 90 min with 300 mM sucrose, 100 mM mannitol, 10 mM Tris-HEPES to reduce the intravesicular space (Inoue et al., Hepatology 2: 572-579, 1982). Uptake was measured in media containing 100 mM mannitol, 10 mM Tris-HEPES, 100 mM NaCl, 0.1 mM unlabelled urate and 0.67 mM [2-^{14}C]urate. Filled circles (n=6) represent the uptake into untreated vesicles in the absence of sucrose, open circles (n=4) represent the uptake into preshrunken vesicles in the presence of 300 mM sucrose.

Further evidence for transmembrane movement of urate was provided by trans-stimulation experiments. BLMV preincubated for 90 min in 2 mM unlabelled urate, 100 mM mannitol and 10 mM Tris-HEPES (pH 7.6) yielded a stimulation of urate uptake of $21.0 \pm 4.5\%$ (n=3 $\pm$ SE) after 1 min of incubation compared to vesicles preincubated in the absence of urate. The final concentration of urate in the transport media was identical.

Urate uptake by hepatopancreatic basolateral membrane vesicles consisted of two different processes, a high-affinity transport process with an apparent K_M-value of 1.04 mM (v_{max} = 29 pmol/s/mg protein; derived from analysis of Eadie-Scatchard plot) and a diffusion process directly dependent on the extravesicular urate concentration.

The relative specificity of urate transport was determined in a series of three to five experiments. Urate analogues and inhibitors were added in a concentration of 1 mM to the uptake medium containing 0.1 mM unlabelled urate and 0.67 mM labelled urate. Unlabelled urate was the strongest inhibitor of [2-^{14}C]urate uptake ($30 \pm 5\%$ inhibition, n=5, $p < 0.01$) after 1 min of incubation while the urate transport inhibitor pyrazinoic acid (Frankfurt SJ & Weinmann EJ, Proc Soc Exp Biol Med 159: 16-20, 1978) and the uricase inhibitor oxonic acid (Fridovich J, J Biol Chem 240: 2491-2494, 1965) were slightly less effective in inhibition ($26 \pm 11\%$, n=3, $p < 0.05$ and $25 \pm 5\%$, n=3, $p < 0.01$, respectively). Hypoxanthine, purine and allantoin had weaker inhibitory effects on urate uptake ($24 \pm 12\%$, $21 \pm 10\%$ and $19 \pm 8\%$, respectively, all n=3 and $p < 0.05$).

In the present study the existence of an urate uptake mechanism in the basolateral membrane of lobster hepatopancreas was demonstrated. Urate uptake consists of a high-affinity transport component and a diffusional component. The uptake shows affinity for different purine derivatives as well as sensitivity for the uricase inhibitor oxonic acid and the urate transport inhibitor pyrazinoic acid indicating that the urate transport mechanism might simultaneously have the ability of degrading urate (Abramson et al., Am J Physiol 242: F158-F170, 1982).

We would like to thank Dr. Ruth Abramson for the generous gift of the [2-^{14}C]urate and for helpful discussions. The work was supported by the Deutsche Forschungsgemeinschaft (Gr 456/12-2).

ORGANIC ANION TRANSPORT MECHANISMS IN RENAL TISSUE FROM Fundulus heteroclitus and Cancer borealis AS PROBED BY EPI-FLUORESCENCE MICROSCOPY and VIDEO IMAGE ANALYSIS

Daniel E. Bowen[1] and David S. Miller[2]
[1]Biology Department, Benedictine College, Atchison KS 66002
[2]Laboratory of Cellular and Molecular Pharmacology, NIH/NIEHS,
Research Triangle Park, NC 27709

The renal proximal tubule plays an major role in the elimination of potentially toxic organic anions, e.g., metabolic wastes, drugs and environmental pollutants and drug and pollutant metabolites. This summer we used epi-fluorescence microscopy and video image analysis to investigate two aspects of the renal organic anion transport mechanism. The first is concerned with uptake at the basolateral membrane, the rate-determining step in secretion. Recent experiments show that organic anion uptake and net secretion can be increased by indirect coupling to the Na gradient (Pritchard and Miller, In: The Kidney: Physiology and Pathophysiology, Raven Press, 1992, pp 2921-2945). This occurs through two energetically coupled events at the basolateral membrane: monovalent-divalent organic anion exchange and Na-coupled divalent anion uptake. Importantly, in the absence of added divalent organic anions, monovalent organic anions are still actively accumulated. Is the same Na-dependent, indirect coupling mechanism used to drive basal transport?

To answer this question, we measured fluorescein (FL) uptake and efflux in proximal tubules from sea water adapted killifish, Fundulus heteroclitus. Recent studies have shown that FL is a substrate for the renal organic anion transport system in rabbit and teleost fish proximal tubules (Sullivan et al, Amer. J. Physiol. 258:F46-F51, 1990; Miller and Pritchard, Amer. J. Physiol. 261:R1470-R1477, 1991). Unlike other commonly studied substrates for transport, which can only be measured using chemical or radioisotopic means, FL is intensely fluorescent and its uptake into the cells and lumens of proximal tubules can be followed using a fluorescence microscope. Thus, use of this compound adds a new dimension to the study of organic anion transport, spatial resolution.

When killifish tubules were incubated in a buffered teleost saline solution (containing, in mM: 140 NaCl, 2.5 KCl, 1.5 $CaCl_2$, 1.0 $MgCl_2$ and 20 TRIS, at pH 8.25) with 1 μM FL, the dye rapidly accumulated in the cells and lumen. Within 15 min, uptake in both compartments had reached a steady state. If 10 μM glutarate was present in the buffer from time-zero tubules showed roughly the same time course of uptake over 20 min, followed by a second uptake phase and a second higher plateau. Parallel patterns of uptake were seen in the tubular lumina where fluorescent intensity was 50-100% higher. Calibration of the imaging system using glass capillaries with known concentrations of FL indicated that steady state concentrations in the cells were about 25 μM at the first plateau and about 40 μM at the second plateau. Thus, even in the absence of added glutarate, cells accumulated FL to a level many times that of the medium. Glutarate caused a delayed increase in accumulation; the delay most likely represents the time needed to accumulate sufficient glutarate to provide an additional driving force for anion exchange.

We have used LiCl as a tool to determine if basal, uphill accumulation of organic anions was driven by indirect coupling to Na. Lithium inhibits coupled organic anion transport by blocking Na-divalent organic anion uptake, but it does not affect anion exchange or disrupt cell metabolism (Pritchard, Amer. J. Physiol. 255:F597-F604, 1988; Miller, unpublished data). Two types of experiment showed that basal FL uptake is driven by indirect coupling to Na. First, when killifish proximal tubules were incubated with buffer containing 1 μM FL (no added glutarate) and increasing concentrations of LiCl, FL accumulation was inhibited in a dose dependent manner; with 20 mM LiCl, inhibition was nearly complete. Second, when tubules were loaded to steady state in buffer with 1 μM FL and then 20 mM LiCl added to the medium (without removing the FL), a net efflux of dye from the tissue was observed. The time course of efflux from the cells showed a 1-2 min delay followed by a single exponential with a mean half-time of 32 min (data from 5 tubules). In these efflux experiments, the buffer used contained no added divalent organic anions. Therefore, the most likely explanation for FL runout is that under basal conditions divalent organic anions, such as α-ketoglutarate (the major divalent organic anion in proximal tubule cells), are produced by intermediary metabolism and leak into the medium possibly in exchange for FL. Normally, these divalents anions are returned to the cells by the Na-dependent cotransporter and the steep divalent anion gradient (in>out) would drive FL uptake via exchange. When Li was added to the medium, divalent organic anion uptake was blocked, reducing the gradient for divalent anions. This, in turn, inhibited FL uptake and caused net loss of FL from loaded tubules.

The second question addressed concerns the distribution of organic anions within renal cells. We recently found that in crab urinary bladder (an invertebrate tissue with transport functions like vertebrate proximal tubule), intact teleost proximal tubules and monolayers of renal cells in culture, FL is distributed over two intracellular compartments; one is diffuse and cytoplasmic, the other is concentrated and vesicular (Miller et al, Bull. MDIBL 30:56-57, 1991; Miller et al, Am. J. Physiol., submitted). To determine if FL was trapped in vesicles as they formed at the plasma membrane or if the dye was taken up from the cytoplasm after transport into the cell, we incubated bladder slices from Cancer borealis in a crab Ringer's solution (containing, in mM: 449 NaCl, 11 KCl, 12.5 $CaCl_2$, 18.5 $MgCl_2$ and 50 TRIS, at pH 7.8) with FL and various effectors of organic anion transport and measured cytoplasmic and vesicular fluorescence at steady state (90-120 min). Incubation with 10 or 50 μM glutarate stimulated 0.5 μM FL uptake in cytoplasm and vesicles (Table 1). Uptake into the cytoplasm and vesicles was reduced by LiCl and the monovalent organic anions, probenecid and p-aminohippurate. Thus, treatments that stimulate transport at the basolateral membrane (low glutarate concentrations) stimulated vesicular uptake and treatments that inhibited transport (Li and competitor organic anions) reduced vesicular uptake. This indicates that vesicular uptake is dependent on the cytoplasmic FL concentration rather than the medium concentration. Thus, accumulation in vesicles is a two step process, involving uptake into the cytoplasm mediated by the basolateral transporter followed by transport into vesicles. This conclusion has been confirmed recently by microinjection experiments with renal cells in culture. Moreover, those studies have shown that uptake by vesicles is concentrative, specific and energy dependent (Miller et al, Am. J. Physiol., submitted). The role that intracellular compartmentation plays in overall organic anion secretion remains to be determined.

Table 1. Effects of modifiers of organic anion transport on cytoplasmic and vesicular levels of fluorescein in C. borealis urinary bladder.

Treatment	Cytoplasm	Vesicles
Control	41 ± 1 (11)	64 ± 3
10 μM Glutarate	138 ± 5 (7)	212 ± 5
50 μM Glutarate	81 ± 3 (9)	107 ± 3
20 mM LiCl + 10 μM Glutarate	32 ± 2 (5)	NVD
1 mM Probenecid	27 ± 2 (6)	NVD
1 mM p-Aminohippurate	25 ± 2 (7)	NVD

Data given as mean ± SE pixel intensity (scale 0-255) for each intracellular region; the number of measurements is in parentheses. NVD, no vesicles detected (punctate areas of high fluorescent intensity were not seen). Because of the contribution of out of focus cytoplasmic fluorescence above and below vesicles and the small vesicle diameter (2-5 μm), measured vesicle intensities underestimate actual values by 5-40 fold.

Supported in part by a Pew Fellowship to D.E. Bowen.

BUMETANIDE UPTAKE IN ISOLATED HEPATOCYTES OF THE LITTLE SKATE (RAJA ERINACEA)

M. Blumrich[1], E. Petzinger[1] and J. L. Boyer[2]
[1]Institut für Pharmakologie und Toxikologie, Justus-Liebig-Universität Giessen, 6300 Giessen, Germany
[2]Department of Medicine and Liver Center, Yale University School of Medicine, New Haven, CT 06510

In dog, rats and men the loop diuretic drug bumetanide is cleared from blood mainly through the liver (Busch et al., Arzneim. Forsch. Drug Res. 29: 315, 1979). Recent studies with isolated and cultured hepatocytes demonstrated sodium coupled carrier mediated bumetanide uptake via a high affinity and a low affinity transport system in isolated rat hepatocytes (Petzinger et al. Am. J. Physiol. 256: G 78, 1989; Föllmann et al. Am. J. Physiol. 258: C 700, 1990). Photoaffinity labeling of these putative transporter proteins identified a 52 - 54 KDa integral protein in rat liver basolateral plasma membranes (Petzinger et al. Eur. J. Pharmacol. Mol. Pharmacol. Sect. 208: 53, 1991). Other loop diuretics and bile acids competed with bumetanide for uptake and photoaffinity labeling suggesting a common uptake mechanism.

In isolated hepatocytes from the elasmobranch Raja erinacea only a single sodium-independent sinusoidal bile salt transport system has been characterized (Fricker et al., Am. J. Physiol. 253: G816, 1987; Smith et al., Am. J. Physiol. 252: G479, 1987). The elasmobranch organic anion transporter is thought to be an archaic transport system which may be present in mammalian species. It was therefore of interest to examine the mechanisms of hepatic uptake of bumetanide in liver cells from Raja erinacea.

Hepatocytes were isolated from male skates by a collagenase perfusion technique (Smith et al., J. Exp. Zool. 241: 291, 1987) and resuspended in elasmobranch Ringers. (^{3}H)-bumetanide uptake was measured by a rapid centrifugation method at 15 °C (Ballatori and Boyer, Am. J. Physiol. 254: R801, 1988). Uptake was linear for at least 60 s enabling initial uptake rates (V_i) to be calculated from this portion by linear regression.

Results indicate that bumetande uptake into isolated hepatocytes was saturable and energy dependent as well as temperature sensitive. However, in contrast to conjugated and unconjugated bile acids, bumetanide uptake was Na^+-dependent (see Table 1). When sodium ions were substituted in the incubation medium with choline, bumetanide uptake (V_i) was reduced to 58 %.

<u>Uptake (V_i) into isolated skate hepatocytes at 15 °C</u>

	TAUROCHOLATE ($V_i \pm SD$)	**CHOLATE** ($V_i \pm SD$)	**BUNETANIDE** ($V_i \pm SD$)
+Na$^+$	20.9 ± 5.2	23.7 ± 5.4	23.7 ± 5.7
-Na$^+$	19.4 ± 4.7	22.6 ± 3.8	13.8 ± 3.8
Inhib.	7 %	5 %	42 %

Table 1. Initial uptake rates (V_i) of (^{3}H)-taurocholate, (^{14}C)-cholate and (^{3}H)-bumetanide in isolated skate hepatocytes 3 x 10^6 hepatocytes/ml suspension were preincubated either in Na$^+$ or in choline elasmobranch Ringer solutions at 15 °C. V_i rates were determined by linear regression for the uptake of 11 nM (^{3}H)-taurocholate/10 μM taurocholate, 1.25 μM (^{14}C)-cholate/5 μM cholate or 60 nM (^{3}H)-bumetanide/7 μM bumetanide from the 15, 45, 75 and 105 seconds values. n = 6, X ± SD

Kinetic experiments (performed with hepatocytes pooled from livers of at least two skates) and resuspended either in Na$^+$ or in choline-elasmobranch Ringer media revealed the following kinetic constants:
- measured in Na$^+$-Ringer: K_m = 33 - 70 μM;
$$V_{max} = 333 - 400 \text{ pmol x mg}^{-1} \text{ x min}^{-1}$$

- measured in choline-Ringer: K_m = 33 - 66 μM;
$$V_{max} = 250 - 250 \text{ pmol x mg}^{-1} \text{ x min}^{-1}$$

- calculated Na$^+$-dependent portion: K_m = 52 - 79 μM;
$$V_{max} = 100 - 200 \text{ pmol x mg}^{-1} \text{ x min}^{-1}$$

As illustrated in Fig. 1 taurocholate competitively inhibited only the sodium-independent component of bumetanide transport, while the same bile acid was a non-competitive inhibitor in Na$^+$-Ringer solution. In turn bumetanide competitively inhibited (^{3}H)-taurocholate uptake (data not shown).

To further characterize the mechanism of bumetanide uptake, experiments were performed with 100 μM furosemide (a sulfamoyl benzoic acid derivative like bumetanide) and two bumetanide analogs PF-3034 (500 μM) and PF-2203 (500 μM). In rat liver cells these analogs almost exclusively inhibited cholate and bumetanide uptake but not sodium-dependent taurocholate uptake (Petzinger et al. Am. J. Physiol. 1992 submitted). In the present study with skate hepatocytes, these compounds inhibited taurocholate uptake but also preferentially the sodium-dependent bumetanide uptake system. Taurocholate (100 μM), cholate (100 μM) as well as probenecid (100 μM) preferentially inhibited sodium-independent bumetanide transport. BSP (25 μM), rose bengal (25 μM), bilirubin (25 μM), digitoxin (25 μM) and DIDS (100 μM) were strong inhibitors of both uptake mechanisms. In contrast aminoisobutyric acid (AIB, 2 mM) and para-aminohippuric acid (PAH, 1 mM) did not inhibit.

In summary, the results indicate that, unlike tauro-cholate transport in the little skate, a sodium-dependent as well as a sodium independent transporter is involved in bume-tanide uptake into hepatocytes from <u>Raja erinacea</u>. The sodium-independent system is shared by bile acids, which is consi-stent with the hypothesis that this transport system is responsible for the uptake of a variety of different xenobio-tics in lower vertebrates whereas the sodium-dependent bile acid transporter has evolved later in evolution. In contrast, the sodium-dependent transport system of the bumetanide uptake has evolved earlier. Therefore the liver cells of this marine species should be an excellent model to study phylogenetic aspects of bumetanide and organic anion transport.

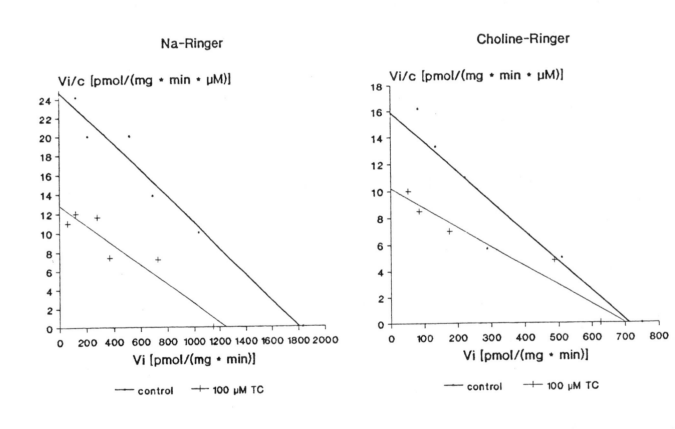

Figure 1. Eadie-Scatchard plot of the bumetanide uptake into freshly isolated skate hepatocytes in the presence of 100 μM taurocholate. In sodium elasmobranch Ringer solution (left figure) taurocholate non-competitively inhibited uptake while competitive inhibition was demonstrated in choline-Ringer (right figure). Each experiment was performed in duplicate with at least three different cell preparations.

Supported by Sonderforschungsbereich 249, project B3 of E. P., Giessen, Germany and DK-34989.

EFFECT OF CADMIUM ON K$^+$-STIMULATED NEUTRAL p-NITROPHENYLPHOSPHATASE IN RECTAL GLAND PLASMA MEMBRANES OF SQUALUS ACANTHIAS

Evamaria Kinne-Saffran, Christiane Pfaff, Marion Hülseweh, and Rolf K.H. Kinne
Max-Planck-Institut für Systemphysiologie
Rheinlanddamm 201, 4600 Dortmund, F.R.G.

In previous studies we have shown that low concentrations of cadmium inhibit the Na$^+$-K$^+$-ATPase activity measured in isolated plasma membranes from shark rectal gland [Kinne-Saffran et al., MDIBL Bull. 26: 15-17, 1986]. In accordance with results obtained in isolated perfused rectal glands [Silva et al., in: 2nd Annual Report of Progress - Center for Membrane Toxicity Studies, pp. 60-62, 1987] it was postulated that cadmium exerts its inhibitory action at the cytoplasmic face of the membranes. Partial reactions of the enzyme taking place at the membrane/cytosol interface are the interaction with sodium, ATP and magnesium. Subsequently performed experiments showed that the sodium site was not a target site for cadmium. Increasing the ATP or the magnesium concentration reduced, however, the inhibitory effect of cadmium on Na$^+$-K$^+$-ATPase activity [Kinne-Saffran et al., MDIBL Bull. 30: 38-40, 1991]. Since ATP is a strong chelator for magnesium and cadmium these studies did not allow to differentiate between a competition of cadmium with the magnesium site of the enzyme proper or with the magnesium-ATP interaction. Therefore, the effect of magnesium and cadmium on a partial reaction of the Na$^+$-K$^+$-ATPase reaction cycle, the K$^+$-stimulated neutral p-nitrophenylphosphatase was investigated in the present study.

Plasma membranes were isolated from rectal glands of Squalus acanthias by differential centrifugation as source for the K$^+$-stimulated neutral p-nitrophenylphosphatase activity [Hannafin et al., J. Comp. Physiol. B 155: 415-421, 1985]. Enzyme activity in lyophilized membranes was determined at 15°C in the presence of 50 mmol imidazole buffer, pH 7.6, 3 mmol p-nitrophenylphosphate, 1 mmol magnesium chloride, with and without 5 mmol potassium chloride as standard incubation medium. The release of p-nitrophenolate was followed at a wavelength of 405 nm in a recording spectrophotometer. The difference between the activity in the presence and absence of K$^+$ was considered to represent the K$^+$-stimulated neutral p-nitrophenylphosphatase related to the Na$^+$-K$^+$-ATPase in these membranes.

Analysis of the effect of cadmium on the K$^+$-stimulated neutral p-nitrophenylphosphatase activity revealed that the inhibition of cadmium increased with time reaching a maximum after 30 minutes of incubation at 15°C. Without preincubation the apparent K_i of cadmium was 8.9 x 10^{-6} mol. The K_i after preincubation of the membranes was 2.2 x 10^{-6} mol, thus a 4-fold increase in sensitivity of the enzyme was observed. It was also found that the degree of inhibition by various cadmium concentrations was strongly dependent on the concentration of magnesium present during the preincubation period. When the dose response curve in the presence of 1 mmol magnesium was compared with the dose response curve in the presence of 10 mmol magnesium a shift to the right was observed, which resulted in a decrease in the apparent K_i in the presence of 10 mmol magnesium to 1.4 x 10^{-5} mol cadmium. This represents a 6-fold decrease in sensitivity of the enzyme against cadmium inhibition (figure 1). Similarly, at a constant cadmium concentration increasing magnesium concentrations from 0.5 mmol to 5.0 mmol led to an attenuation of the inhibition from 40% to only 18%.

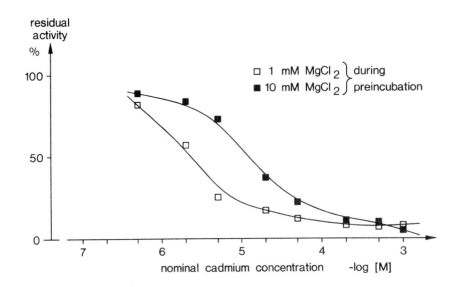

Figure 1. Effect of magnesium on cadmium-dependent inhibition of K[+]-stimulated neutral p-nitrophenylphosphatase. Lyophilized rectal gland plasma membranes were incubated for 30 min at 15°C at the cadmium and magnesium concentrations indicated in the figure and the enzyme activity was subsequently assayed in the presence of 1 mM or 10 mM magnesium. Mean values from 3 determinations are given. Nominal concentrations have been calculated according to the amount of cadmium chloride added to the assay medium.

In order to further investigate the mode of action of cadmium on the K[+]-stimulated neutral p-nitrophenylphosphatase we studied the effect of cadmium on the affinity of the enzyme towards magnesium. In the absence of cadmium a K_m for magnesium of 0.88 ± 0.29 mmol was found. In the presence of 4×10^{-5} mol cadmium the K_m value doubled to 1.73 ± 0.33 mmol, whereas the maximum velocity was only reduced by about 25% (167 ± 32 μmol/min vs. 140 ± 16 μmol/min; n = 3).

These studies suggest that one of the sites at which cadmium interacts with the Na[+]-K[+]-ATPase in shark rectal gland plasma membranes is a magnesium binding site which plays an essential role in initiating ATP hydrolysis. Since it is assumed that these magnesium sites are located at the cytoplasmic face of the enzyme molecule the intracellular magnesium concentration may play an important role in modulating the cellular toxicity of cadmium. Thus, high intracellular magnesium concentrations could exert a protective effect against the toxicity of cadmium in renal cells.

Supported by grant NIEHS ES 03828

CLONING AND CHARACTERIZATION OF SEQUENCES ENCODING THE MAJOR ISOFORM OF NA,K-ATPase IN SHARK RECTAL GLAND

Edward J. Benz, Jr.[1], Paul Scofield[2], James Z. Appel[1]
Jennifer Benz[1], Cynthia Boyd[1], Jonathan Chai[1],
Prasad Devarajan[1], Leonard Su[1]
[1]Department of Internal Medicine, Hematology Section
Yale University School of Medicine
333 Cedar Street, P.O. Box 3333
New Haven, CT 06510-8056
[2]Cambridge University, Cambridge, England

We and others (Matsuda, T. et al., J. Biol. Chem. 259:3858-3863, 1984; Sweadner, K.J., et al., Biochim Biophys. Acta 988:185-220, 1989; Orlowski, J., et al., J. Biol. Chem., 263:17817-17821, 1988; Orlowski, J., et al., J. Biol. Chem., 263:10436-10442, 1988; Schneider, J.W., et al., Proc. Natl. Acad. Sci., USA, 85:284-288, 1988) have previously demonstrated that 3 isoforms of the α subunit of the Na,K-ATPase exist in mammalian and avian species. However, only one isoform is present in invertebrates (Varadi, A. et al. FEBS Letters, 258:203, 1989). This isoform most closely resembles the A3, or neural, isoform of mammals and birds (Schneider, J.W., et al., Proc. Natl. Acad. Sci, USA, 85:284-288, 1988; Varadi, A. et al., FEBS Letters, 258:203, 1989), on the basis of nucleic acid and amino acid sequence homology. The A1 isoform predominates in epithelial tissues, while A2 is especially abundant in muscle. We have attempted to define the number and types of isoforms of Na,K-ATPase present in the rectal gland of the common dogfish shark, Squalus acanthus. We focused our studies upon the rectal gland for two reasons. First, the behavior the Na,K-ATPase has been extensively studied in this tissue; it would be of considerable interest to know whether the Na,K-ATPase activity of the rectal gland arose from a single isoform, or several. Second, if multiple isoforms exist, it would be of interest to know whether the transport epithelium of the rectal gland expresses the A1 isoform, because A1 is the only isoform expressed in epithelial cells of mammals and birds. This information would aid efforts to discern a functional basis for the existence of multiple isoforms.

In order to examine the isoforms present in the dogfish rectal gland, we elected to clone their cDNAs. We adapted methods, based upon the polymerase chain reaction (PCR), that we had employed to characterize Na,K-ATPase cDNAs in the common fruitfly, Drosophila melanogaster (Varadi, A. et al., FEBS Letters, 258:203, 1989). Briefly, messenger RNA was isolated from six rectal glands by use of standard commercial RNA extraction kits. The mRNA was reverse transcribed into cDNA with reverse transcriptase. Selective amplification of portions of the cDNA moieties that were complementary to the Na,K-ATPase mRNA was then accomplished by PCR, using oligonucleotide primers complementary to highly conserved regions of Na,K-ATPase mRNA. The primers were chosen by comparing previously cloned cDNA sequences from many species: rat, Torpedo, chicken, sheep, pig, human, and Drosophila (Varadi, A. et al., FEBS Letters, 258:203, 1989). Two extremely conserved regions of homology emerged from this comparison: amino acids within the phosphorylation site (amino acids 368-376) and a stretch of amino acids within the FITC site (amino acids 502-510). The former site was used to design a "sense" strand oligonucleotide primer, 23 bases in length, while the latter site was utilized for design of the "anti-sense" strand primer, also 23 bases in length. The region between these two primers was then selectively amplified by PCR, using temperatures of $95°C$ for one minute (denaturation), $55°$ for 2 minutes (annealing), and $72\,°C$ for 2.5 minutes (chain extension). The DNA products from this reaction were then analyzed by agrose gel electrophoresis. A single DNA band, 429 bases in length, was observed. This result was consistent with selective amplification of cDNA sequences located between the two primer sites, which are 130-145 amino acids apart in the Na,K-ATPase's of many species; it suggested strongly that the band represented a portion of the shark rectal gland Na,K-ATPase cDNA.

The 429 base pair band was recovered by subcloning into a standard bacterial plasmid vector, and subjected to DNA sequencing using standard techniques (Varadi, A., et al., FEBS Letters.258:203, 1989). The sequence analysis revealed that the amplified cDNA segment encoded an Na,K-ATPase isoform highly homologous to the A3 isoform. The amino acid sequence encoded by this cDNA segment exhibited 89% homology to the rat and human A3 isoform; homologies for the A1 and A2 isoforms were significant, but considerably lower: 76 and 73%, respectively. The shark sequence also matched the A3 sequence at several positions where rat and human A3 amino acid sequences diverge from their A1 counterparts. This is most apparent in the regions encoding amino acids 489-498, where A1 and A3 isoforms differ significantly. Within this region, the shark sequence exhibits a 9/10 sequence match with the A3, but a 0/10 match with the A1 sequence. These findings strongly suggest that the shark sequence is most closely related to the A3 isoform of mammals.

The amplified cDNA segment was also utilized as a probe for RNA analysis, employing stringent Northern blotting hybridization techniques that eliminate cross-hybridization among mRNAs encoding different isoforms (Schneider, J.W., et al., Proc. Natl. Acad. Sci, USA, 85:284-288, 1988). Under these conditions, the probe detected a prominent mRNA band in shark rectal gland RNA; the RNA band was 3,800 bases long, consistent with the length of mRNAs encoding Na,K-ATP'ase alpha subunits from all species previously examined (Orlowski, J., et al., J. Biol. Chem., 263:17817-17821, 1988; Orlowski, J., et al., J. Biol. Chem., 263:10436-10442, 1988; Schneider, J.W., et al., Proc. Natl. Acad. Sci, USA, 85:284-288, 1988; Varadi, A., et al., FEBS Letters, 258:203, 1989). The mRNA was present, but less abundant, in other tissues.

Our preliminary results suggest, but do not prove, that the predominant Na,K-ATPase isoform expressed in the dogfish shark is the A3 form, since we found this form rather than A1 in transport epithelium. This pattern is much more typical of that seen in invertebrates, where only one isoform is produced. In mammals, the isoform in transport epithelium is almost always exclusively the A1 isoform. Further studies will be necessary to determine whether other shark isoforms exist and their tissue distributions. This initial result, however, is striking because it identified an A3 isoform in epithelial cells.

Supported by grants from the American Heart Association (Connecticut and Maine affiliates).

PHYSIOLOGICAL AND BIOCHEMICAL EVIDENCE FOR H^+/K^+ ATPase MEDIATED RENAL ACID SECRETION IN THE ELASMOBRANCH

Erik R. Swenson[1], Andrew D. Fine[2], Thomas H. Maren[2],
Evamaria Kinne-Saffran[3], and Rolf K. H. Kinne[3]

[1]Department of Medicine, VA Medical Center and University of Washington School of Medicine, Seattle, WA, 98108, USA;
[2]Department of Pharmacology and Therapeutics, University of Florida College of Medicine, Gainesville, FL 32610, USA;
[3]Max-Planck-Institut fur Systemphysiologie, Dortmund, FRG.

The kidneys of marine elasmobranchs lack carbonic anhydrase (CA) (Hodler et al., Am J Physiol, 183:155,1955). Despite this, they consistently acidify the urine under a variety of conditions (Swenson and Maren, Am J Physiol. 250:F288,1986). During high rates of stimulated acid secretion and HCO3- reabsorption, less than 10% of H^+ formation or OH^- dissipation can be accounted for by the known uncatalyzed reaction rates of CO_2 in the renal tubule (Swenson and Maren, ibid). It has been our goal to elucidate the mechanisms of CA and CO_2 independent H^+ secretion and HCO_3^- reabsorption in marine fish. Insights into these processes may lead to a better understanding of acid-base transport in the mammalian kidney under conditions such as CA inhibition and in segments of the nephron which lack the enzyme.

Little is known about the actual membrane events involved in elasmobranch renal acid secretion. Recently Bevan et al. (J Comp Physiol B. 159:339,1989) demonstrated that renal proximal brush border membrane vesicles of the dogfish are capable of Na^+/H^+ exchange. This process is saturable and inhibited by amiloride. However, direct evidence supporting a role for a luminal membrane Na^+/H^+ antiporter in acid secretion and HCO_3^- reabsorption is lacking. Amiloride did not alter luminal pH in micropuncture studies (Sibernagl et al.,Bull MDIBL 26:156,1986) or reduce urinary acid output in whole animal studies (Swenson, unpublished data). These lack of effects may, however, simply represent the weak inhibitory effect of amiloride in the presence of very high physiologic $[Na^+]$.

Another membrane transporter of possible importance is an apical membrane associated proton translocating ATPase. An attractive candidate is H^+/K^+ ATPase once thought only limited to the gastric mucosa, but now demonstrated in many acid secreting epithelia; mammalian kidney (Wingo, J Clin Invest. 84:361,1989), colon (Perrone and McBride, Pflugers Arch. 416:632, 1990), and lung (Boyd et al., J Physiol. 325:46P,1991), amphibian kidney (Planelles et al., Am J Physiol. 260:F806,1991), and jejunum (White, Am J Physiol. 248:G256,1985) and turtle urinary bladder (Sabatini et al., Clin Res 38:984A,1991). In these tissues, the enzyme is inhibited by SCH 28080, a highly specific inhibitor of mammalian gastric H^+/K^+ ATPase. Therefore, we set out to determine whether renal acid secretion in the elasmobranch involves an H^+/K^+ ATPase.

Spiny dogfish, Squalus acanthias (males, weight range 1.8 - 2.2 kg) were studied 12 to 16 hours after transfer into small plexiglass tanks and placement of caudal artery and urinary papilla catheters. The effects of SCH 28080 were studied under two conditions: a) unstimulated basal acid secretion and b) stimulated acid secretion. In the first group, a continuous infusion of elasmobranch Ringer's solution was begun at 15 ml/h-kg. Thereafter hourly urine samples were collected and measured for pH, volume and titratable acid (TA) concentration. Arterial blood gas samples were also analyzed for pH and PO_2 and total CO_2 content at two hour intervals to document stable oxygenation and acid base status. After two hours, the fish were either given SCH 28080 (62 mg/kg) disolved in 5 ml of ethanol (n=5) or 5 ml of ethanol (n=4) as a control. These were given over 30 minutes by a constant infusion pump and measurements were continued over the next 3 hours. In the second group, renal acid secretion was stimulated by a constant infusion of 225 mM imidazole in Ringer's at 7.5 ml/h-kg, which we have shown previously to increase acid output 5-10 fold (Swenson and Maren,ibid). After three hours, either 62 mg/kg SCH 28080 dissolved in ethanol (n=5) or ethanol alone (n=5) were given as described above. The dose of SCH 28080 was calculated to achieve a concentration of roughly 0.5 - 1.0 mM if distributed into total body water.

Luminal membrane fractions from dogfish kidney were prepared by a modified calcium precipitation method (Kinne-Saffran et al., Bull MDIBL 24:61,1984). Alkaline phosphatase, a brush border membranemarker, was enriched approximately 12 fold whereas Na^+/K^+ ATPase, a basolateral membrane marker was reduced 0.66. H^+/K^+ ATPase activity in these membranes was measured as the rate of K^+ stimulated ATP hydrolysis in the presence of ouabain. The assay conditions were as follows: 20 mM HEPES, 3 mM Tris-ATP, 2 mM ouabain at pH 7.0 and varying amounts of KCl from 10 -100 mM. ATP hydrolysis was determined by the release of inorganic phosphate (Pi). Incubation time was 30 minutes at 25 degrees C. Several cations were tested to examine the specificity of K^+ in stimulating this ATPase activity. The inhibitory effects of SCH 28080 on the stimulation by 75 mM KCl were studied at 0.5 mM.

In unstimulated fish, SCH 28080 increased urinary pH from 5.81 to 6.25; reduced urine flow from 0.9 to 0.3 ml/h-kg and reduced urinary [TA] from 34 to 16 mEq/l. Urinary pH, flow and [TA] in the control (ethanol alone) fish were not altered. Figures 1 and 2 show equally marked reductions (87% and 77%, respectively) in the rate of titratable acid secretion with SCH 28080 in unstimulated and imidazole stimulated fish.

In the absence of K^+, the renal plasma membrane fractions had an average ATPase activity in the presence of Mg^{++} of 6.8 ± 1.2 umol Pi/h-mg protein. Potassium at: 10 mM increased this activity by 13.4 ± 6.3%, at 50 mM by 27.9 ± 5.5%, at 75 mM by 32.0 ± 8.3 % and at 100 mM by 38.6 ± 8.3% (means ± SD). This concentration dependent response yields an apparent Km of roughly 25 mM. At 75 mM, RbCl caused a 38% increase, CsCl a 33% increase and LiCl a 26% increase in ATPase activity. SCH 28080 at 0.5 mM caused a 55% inhibition of the ATPase activity stimulated by 75 mM KCl.

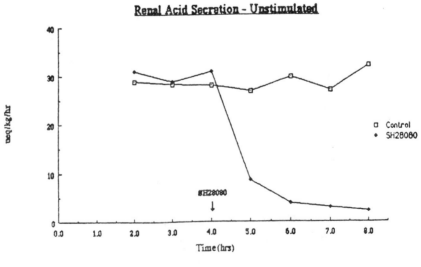

FIGURE 1

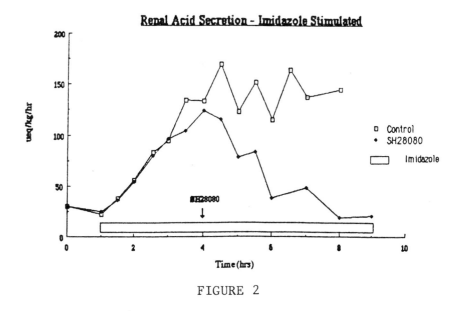

FIGURE 2

For the first time, it is possible to inhibit acid secretion in the marine fish. All previous attempts have failed including those known to decrease mammalian renal acid output such as metabolic alkalosis, CA inhibitors and amiloride. Reduction of total titratable acid output occurs as a consequence of raised pH (lower [TA]) and reduced flow. We did not measure urinary phosphate, so it is not possible to determine whether inhibition of acid output in the unstimulated state is solely an effect of reduced phosphate secretion. Under other conditions (such as phosphate loading) it appears tightly linked to acid output and could be conceived to be primary with H^+ following secondarily. However, the increase in urinary pH and reduction in stimulated H^+ output with imidazole (when it becomes the predominant urinary buffer) are more consistent with a direct SCH 28080 effect on H^+ secretion. These physiological data are supported by our biochemical studies revealing a brush border membrane K^+ stimulated, ouabain insensitive, ATPase activity reduced by a known inhibitor of H^+/K^+ ATPase. It is clear, however, that this putative enzyme differs from that in the mammal and amphibian. H^+/K^+ ATPases from these vertebrate classes, have a lower Km for K^+ (0.5-1.0 mM) and are more sensitive to K^+ than other univalent cations.

Our findings point to a major role of renal H^+/K^+ ATPase in acid secretion in marine elasmobranchs. It is conceivable that the H^+/K^+ ATPase we have demonstrated in the shark developed early in vertebrate evolution and that selectivity and sensitivity for K^+ arose later. Inhibition by SCH 28080 and preliminary evidence that multiple segments of the skate nephron stain specifically along the apical (luminal) membrane with antibodies against two portions of the alpha chain of hog gastric H^+/K^+ ATPase (Swenson et al, FASEB J 1992, in press) suggest that despite these differences some features of this proton translocating ATPase have been preserved over time. Several aspects of elasmobranch renal function and acid-base regulation are possibly explained by H^+/K^+ ATPase mediated acid secretion. The first is that the enzyme generates a proton for transport directly in the hydrolysis of ATP, thus eliminating any need for CA and CO_2. Furthermore, since we have discovered a means to inhibit renal H^+ secretion, it will be possible to test directly whether bicarbonate reabsorption is mediated via H^+ secretion or by some independent mechanism of direct ionic transport. Lastly, H^+/K^+ ATPase mediated acid secretion may explain the tight linkage of acid and phosphate excretion and provide a mechanism for potassium reabsorption in the elasmobranch kidney, a process surprisingly neglected in fish renal physiology.

This work was supported by NIH grant # EY02227 to THM.

STOICHIOMETRY OF SODIUM-D-GLUCOSE COTRANSPORT
IN RENAL BRUSH BORDER MEMBRANES OF THE
DOGFISH (SQUALUS ACANTHIAS) AND THE LITTLE SKATE (RAJA ERINACEA)

Rolf K.H. Kinne, Evamaria Kinne-Saffran, Marion Hülseweh
Max-Planck-Institut für Systemphysiologie
Rheinlanddamm 201, 4600 Dortmund, F.R.G.

Previously we have demonstrated the presence of sodium-D-glucose cotransport in isolated renal brush border membrane vesicles from hagfish, dogfish, and flounder [Kinne, Issues Biomed. 15: 69-94, 1991]. These studies were now extended to the skate in order to provide functional information on glucose transport in both species in parallel to investigations at the mRNA and DNA level.

Brush border membranes were isolated from the caudal two thirds of the dogfish kidney and whole kidneys of the skate as described previously [Bevan et al., J. Comp. Physiol. B 159: 339-347, 1989]. Transport measurements were performed by a rapid filtration method at 15°C. To determine the stoichiometry of the sodium-D-glucose cotransport two methods were employed. One is based on the analysis of the number of sodium ions interacting with the transport system when measuring the sodium dependence of the initial rate of D-glucose uptake ("activation method"). The other determines those sodium gradients which, at a given ratio of intravesicular to extravesicular D-glucose, prevent the efflux of the sugar from the vesicles ("thermodynamical analysis") [for further information see Turner, Ann. N.Y. Acad. Sci. 456: 10-25, 1985].

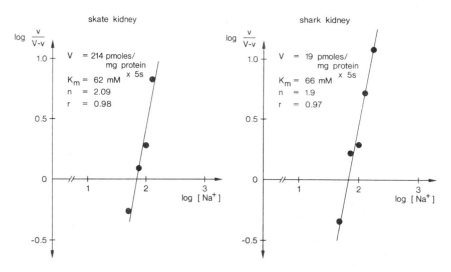

Figure 1. Stoichiometry of the sodium-D-glucose transporter in skate and shark kidney brush border membranes (activation method). Uptake of D-glucose (0.1 mM) was measured for 5 s in media containing various concentrations of NaCl. NaCl was replaced isoosmotically by choline chloride. The vesicles were electrically short-circuited by the presence of 100 mM KSCN at both inside and outside and contained 15 µg valinomycin/mg protein. Sodium-dependent D-glucose transport was calculated as the difference between the uptake in the presence and absence of sodium. The data points represent mean values of 3 experiments performed as duplicates. The standard deviation of the mean values was ± 9% or less.

In figure 1 the sodium dependencies of D-glucose uptake into renal brush border vesicles isolated from shark kidney and skate kidney are compared. It can be seen in the Hill plot that both sodium-D-glucose cotransport systems exhibit an almost identical sodium dependence with a halfsaturation between 62 and 66 mM and a slope of 1.9 and 2.09, respectively. These data suggest that both transport systems interact with two sodium ions and accordingly should exhibit a 2:1 stoichiometry during transport.

This hypothesis was investigated further by preloading brush border membrane vesicles with D-glucose and sodium and measuring efflux of D-glucose into media containing different sodium concentrations. The results of these studies are compiled in table 1.

Table 1.
Changes in intravesicular D-glucose content
after dilution into media varying in sodium concentration

			glucose content after 4 s		
D-glucose gradient glu_i/glu_o	sodium gradient Na_i/Na_o	stoichio-metry at which no flux should occur	shark kidney brush border	skate kidney brush border	p
5/1	1/2.5	1.75	92.8±4.0%	96.1±4.3%	n.s.
5/1	1/5	1	95.4±3.9%	109.1±5.1%	<0.02
5/1	1/10	0.7	104.7±5.4%	128.3±8.5%	<0.02

The intravesicular glucose concentration was 0.1 mM, the intravesicular NaCl concentration 20 mM. NaCl in the efflux media was replaced isoosmotically by choline chloride. The vesicles were electrically short-circuited by 100 mM KSCN and 15 μg valinomycin/mg protein. Values are given as % of the initial content and represent means ± SD of 3 experiments performed in triplicate. The column stoichiometry denotes that stoichiometry at which according to the driving forces employed in this experimental condition no net movement of glucose should occur. Loss of D-glucose suggests a lower stoichiometry, uptake of D-glucose a higher stoichiometry. p-values calculated according to Student's t-test apply to the differences between shark and skate kidney.

It is obvious that under these experimental conditions the brush border membranes of the two species behave differently. In shark kidney brush border membranes efflux of D-glucose still occurs when the D-glucose and the sodium gradient compensate each other - as predicted for a system with a 1:1 stoichiometry - and is only stopped by an even higher inwardly-directed sodium gradient. This result suggests that the stoichiometry of the shark kidney sodium-D-glucose cotransport is between 0.7 and 1 when the energetics of transport are investigated. On the opposite, in skate brush border membranes the transition point between efflux and uptake of D-glucose is clearly above a stoichiometry of one, in agreement with the result obtained from investigating the sodium dependence of transport.

These studies indicate that stoichiometries determined by kinetic or thermodynamical methods do not always yield similar results. One possible explanation for this discrepancy is the presence of internal leak pathways, i.e. in which D-glucose transport can occur without the coupling to sodium [Centelles et al., Biochim. Biophys. Acta 1065: 239-249, 1991]. The apparent stoichiometry as determined thermodynamically would decrease when the relative contribution of such leak pathway increases. From the studies reported above it could be hypothesized that the skate sodium-D-glucose transport system has less internal leak pathways than the transport system in the shark. Since, thus far, in these two species no differences between the transport systems have been dectected at the molecular level [Shetlar et al., MDIBL Bull. 30: 35-37, 1991], this might suggest that different degrees of coupling can occur within the transport molecule or between the transport molecules in the tetrameric configuration which has been found to represent the minimum functional unit for sodium-coupled D-glucose transport across mammalian membranes [Lin et al., Biochim. Biophys. Acta 777: 201-208, 1984].

Funded by grants of the Max-Planck Society to R.K. and E.K.-S.

SEQUENCE COMPARISON OF THE SODIUM-D-GLUCOSE COTRANSPORT SYSTEM IN A VARIETY OF AQUATIC ORGANISMS.

Alison Morrison-Shetlar[1], Richard Moore[2], Beate Schölermann[1], Daniel Kinne and Robert Shetlar[1],

[1]Max-Planck Institut für Systemphysiologie, 4600 Dortmund 1, F.R.G. and [2]Hamilton College, Clinton, NY 13323.

The sodium-D-glucose cotransporter is an important constituent in the transfer of glucose across the brush border membranes of renal and intestinal epithelia. We have isolated, from rabbit cortex, the full cDNA sequence coding for this transporter [Morrison et al, BBA, 1089:121-123, 1991] and used this information to probe DNA and RNA isolated from a variety of marine organisms in an attempt to find sequence homology between species.

RNA and DNA were isolated from Atlantic hagfish (Myxine glutinosa), bloodworm (Glycera dibranchiata), little skate (Raja erinacea), spiny dogfish (Squalus acanthias), winter flounder (Peseudopleuronectes americanus), toadfish (Opsanus tau), goosefish (Lophius americanus) and crab (Carcinus maenas)

RNA was isolated from kidney cells as described by Stallcup and Washington, [J.Biol Chem. 258:2802, 1983]. The RNA was transcribed into cDNA by the method of Okayama and Berg, [Mol.Cell.Biol, 2:161, 1982] using the GeneAmp RNA PCR Kit (Perkin Elmer Cetus). DNA was isolated from muscle tissue by the method of Gross-Bellard, Oudet and Chambon, [Eur.J,Biochem, 36:32, 1973]. The Polymerase Chain Reaction (PCR) was carried out in the Perkin Elmer PCR 480 Cycler using the GeneAmp PCR System (Perkin Elmer Cetus). Sections of the cDNA/DNA were amplified using selected oligonucleotide primers designed according to the rabbit cortical sequence. Three pairs of primers (Fig 1, segments 1, 2 and 3) were used to obtain the complete sequence from each DNA or RNA pool.

The PCR products were separated on agarose gels and southern blotted [Southern, J.Mol.Biol 98:503, 1975] onto Nytran (Schleicher and Schuell). Sodium-D-glucose cotransport cDNA was labelled by nick translation using biotin labelled-dUTP and Southern blots were screened using the Photogene non-radioactive detection system (BRL Life technologies). The results from gel analysis and southern blots are shown in Table 1.

Fig 1. Diagrammatic representation of the full length sequence coding for the rabbit renal sodium-D-glucose cotransporter. Sections 1,2 and 3 indicate the overlapping PCR products synthesized using specific primers.

The results indicate that most animals tested have the sodium-D-glucose cotransport gene present in the DNA and that it is transcribed into mRNA as indicated by the mRNA-cDNA results (Table 1). The results correspond in general well to studies carried out on the transport system at a functional level. It is known for example that the bloodworm has no sodium-D-glucose cotransport system and was also found to be negative in this study. Of particular interest are the results obtained from the toadfish. It is evident that the DNA contains the sequence for the cotransporter and that it is present in the renal mRNA, however part of the sequence would appear to differ (Table 1, see segment 1) and it is known from transport studies that the toadfish kidney has an extremely low sodium-D-glucose uptake capacity. The goosefish is closely related of the toadfish and the same results were obtained.

Animal	DNA			Renal mRNA			Na^+-D-glucose cotransport
	1	2	3	1	2	3	
hagfish	+	+	+	+	+	+	+
bloodworm	-	-	-	-	-	-	-
skate	+	+	+	+	+	+	+
dogfish	+	+	+	+	+	+	+
flounder	+	+	+	+	+	+	+
toadfish	+	+	+	-	+	+	-
goosefish	+	+	+	-	+	+	-
crab	+	-	+	nd	nd	nd	nd
rabbit	+	+	+	+	+	+	+
mouse	+	+	+	+	+	+	+

<u>Table 1</u> Results of PCR product analysis by gel electrophoresis and Southern blot analysis. + = PCR product obtained and positive by Southern blot analysis and transport. - = no product or transport obtained. nd = not determined.

The PCR products are at present being sequenced to determine if there are indeed differences in base sequence. Additionally, cDNA libraries are being produced from each of the animals to screen for the complete sodium-D-glucose cotransporter sequence and to determine the presence of regulatory sequences.

Acknowledgments: This study was kindly supported by a Pew Fellowship to Richard Moore and a Blum-Halsey Award to Alison Morrison-Shetlar and travel money from Max-Planck Institute for Alison Morrison-Shetlar, Robert Shetlar and Beate Schölermann.

POTASSIUM CHANNELS DO NOT MEDIATE THE INHIBITORY EFFECT OF SOMATOSTATIN AND NEUROPEPTIDE Y IN THE RECTAL GLAND OF <u>SQUALUS ACANTHIAS</u>

Silva, Patricio,[1] Heather Brignull, Judd Landsberg, Hadley Solomon, Douglas Wolff, Richard Solomon[1] and Franklin H. Epstein[2]. Departments of Medicine, Harvard Medical School , New England Deaconess Hospital, Joslin Diabetes Center[1] and Beth Israel Hospital,[2] Boston, Massachusetts, 02215.

Sulfonylureas such as gliburide block the ATP dependent potassium channel. This causes a depolarization of the cell with an attendant increase in cytoplasmic calcium and the release of peptides and neurotransmitters. Hyperpolarizing vasodilators, like lemakalim, open the ATP dependent potassium channels thereby hyperpolarizing the cell.   Calcium activated potassium channels, on the other hand, can be blocked by tetraethylammonium or charybdotoxin.   The purpose of the present experiments was to determine the role of ATP and calcium dependent potassium channels in the inhibitory effect of somatostatin and neuropeptide Y (NPY), peptides that inhibit chloride secretion by the rectal gland. Gliburide was used to block the ATP dependent potassium channels and was expected to prevent the effect of the peptide inhibitors if they work through this type of channels. Lemakalim was used to open them and was expected to mimic the effect of the inhibitors. TEA and charybdotoxin were used to block the calcium dependent potassium channel and were expected to block the effect of the peptide inhibitors if these are the channels that mediate their effect.

Isolated shark rectal glands were perfused using a technique developed in our laboratory (Methods in Ezymology, 192:754-66, 1990). Dogfish were pithed and the rectal glands removed by an abdominal incision. The rectal gland artery, vein and duct were catheterized and the glands placed in a glass perfusion chamber maintained at a temperature of $\pm15^\circ$ C with running sea water. The glands were perfused by gravity at a pressure of 40 mm Hg. The composition of the perfusate was (in mM): Na, 280; Cl, 280; K, 5; bicarbonate, 8; phosphate, 1; Ca, 2.5; Mg, 1; sulfate, 0.5; urea, 350; glucose, 5; ph, 7.6 when gassed with 99% O_2/ 1% CO_2. Rectal gland secretion was collected in tared 1.5 ml centrifuge tubes over 10 minute intervals. Chloride concentration in the rectal gland secretion was measured by amperometric titration. All experiments were performed in such a way that one or more glands received the potassium channel blocker/opener and a parallel experiment(s) was used as a control to decrease differences in the response to the peptide inhibitors related to time and date. The agents that block potassium channels were added to the perfusate at the beginning of the perfusion. Perfusions usually lasted 90 minutes with an initial control period of 30 minutes, an experimental period of also 30 minutes during which the inhibitory peptides, somatostatin and NPY, and lemakalim were perfused, and a final control period of 30 minutes. Statistical analysis was done by unpaired Student's "t" test. Values are mean $\pm$ SEM.

Glyburide, at a concentration of 10^{-5}M and 10^{-4}M, had no effect on chloride secretion in either unstimulated glands or glands stimulated with dibutyryl cyclic AMP 5 x 10^{-5}M and theophylline 2.5 x 10^{-4}M. Glyburide at a concentration of 10^{-4}M did not prevent the inhibitory effect of NPY 10^{-7}M (59%±10% inhibition in the presence of glyburide, n=7, and 68%±7% in its absence, n=8, p=NS) or that of somatostatin 5 x 10^{-7}M (76%±7% inhibition in the presence of glyburide, n=4, and 68%±6% in its absence, n=4, p=NS).

Lemakalim at a concentration of 10^{-6}M, n=4, or 10^{-5}M, n=4, had no effect on chloride secretion by glands stimulated with dibutyryl cyclic AMP 5 x 10^{-5}M and theophylline 2.5 x 10^{-4}M.

Tetraethylammonium 10^{-3}M did not alter the inhibitory effect of somatostatin (45%±12% inhibition in the presence of TEA, n=8, and 36%±12% in its absence, n=6, p=NS) or that of NPY (36%±12% inhibition in the presence of TEA, n=10, and 52%±10% in its absence, n=6, p=NS).

Charybdotoxin did not inhibit the effect of somatostatin in two glands (38% inhibition in the presence of charybdotoxin and 40%±12% in its absence, p=NS).

The observation that glyburide does not prevent the inhibitory effect of either somatostatin or NPY suggests that neither of these peptide inhibitors exert their inhibitory effect through the ATP dependent potassium channel. There is no knowledge about the presence of an ATP dependent potassium channel in the rectal gland cells. If such a channel is not present glyburide would not be expected to have an effect. The additional finding that lemakalim, that opens the ATP dependent potassium channel, had no inhibitory effect suggests that such channels are not present in the rectal gland cells or are not numerous enough to have a measurable effect. Similarly, the failure of TEA and charybdotoxin to prevent the inhibitory effect of somatostatin or NPY suggests that these inhibitory peptide do not act via the calcium dependent potassium channel. These channels are believed to be widely distributed in tissues and are conceivably also present in the rectal gland cells.

Supported by grants NIH DK 18078, and NIEHS ESO 3828 and the Pew, W.R. Hearst Foundation and Burroughs-Welcome Foundation grants to Heather Brignull, Sonya Hornung and Douglas Wolff.

EFFECT OF CADMIUM ON SOMATOSTATIN INHIBITION OF CHLORIDE SECRETION BY THE RECTAL GLAND OF <u>SQUALUS ACANTHIAS</u>

Patricio Silva,[1] Richard Solomon,[1] Heather Brignull, Sonya Hornung, Judd Landsberg, Hadley Solomon, Douglas Wolff, and Franklin H. Epstein.[2] Department of Medicine, Harvard Medical School and New England Deaconess Hospital and Joslin Diabetes Center[1] and Beth Israel Hospital,[2] Boston, MA 02215

Cadmium is known to block calcium channels in certain tissues and to inhibit neurotransmitter release. In previous experiments we found that cadmium blocks the inhibitory effect of neuropeptide Y on forskolin-stimulated chloride secretion and the stimulatory effect of atrial natriuretic peptide. However, ·cadmium did not alter the inhibitory effect of somatostatin (Fifth annual report of progress, CMTS, MDIBL, Sep. 30, 1990, pp30-31). On the same annual report Forrest et al. report that cadmium, at concentrations as low as 5 and 10 µM, reversed the inhibitory effect of somatostatin (Fifth annual report of progress, CMTS, MDIBL, Sep. 30, 1990, pp20-22). In the present series of experiments we reexplored the issue of the inhibitory effect of cadmium on somatostatin.

Isolated shark rectal glands were perfused using a technique developed in our laboratory. Dogfish were pithed and the rectal glands removed by an abdominal incision. The rectal gland artery, vein and duct were catheterized and the glands placed in a glass perfusion chamber maintained at a temperature of 15° C with running sea water. The glands were perfused by gravity at a pressure of 40 mm Hg. The composition of the perfusate was (in mM): Na, 280; Cl, 280; K, 5; bicarbonate, 8; phosphate, 1; Ca, 2.5; Mg, 1; sulfate, 0.5; urea, 350; glucose, 5; pH, 7.6 when gassed with 99% O_2/ 1% CO_2. Rectal gland secretion was collected in tared 1.5 ml centrifuge tubes over 10 minute intervals. Chloride concentration in the rectal gland secretion was measured by amperometric titration.

The glands were perfused with and without cadmium chloride (250 µM). Three intrarterial 1 µg boluses of vasoactive intestinal peptide (VIP) were given sequentially. Somatostatin (5 x 10^{-7}M) was infused for 40 minutes. The second VIP bolus.was given ten minutes after starting the somatostatin infusion.

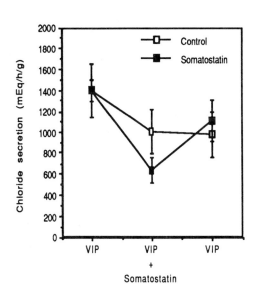

Figure 1. Effect of cadmium chloride on the inhibition by somatostatin of VIP-stimulated chloride secretion. Somatostatin inhibited chloride by 54%, significantly more than the decline in stimulation observed spontaneously in the control group, p < 0.05. Control (open squares), n=6, somatostatin (closed squares), n=9. Values are mean ± SEM.

Somatostatin inhibits VIP-stimulated chloride secretion in the presence, Figure 1, or absence, Figure 2, of cadmium 250 µM. The present experiment differs from our previous experiment in that in the latter, after the rate of secretion had stabilized at a basal level, the glands received forskolin and somatostatin simultaneously. Somatostatin prevented the stimulatory effect of forskolin both in the presence and absence of cadmium. When somatostatin was discontinued secretion rose in response to forskolin. An inhibitory effect of cadmium on chloride secretion was noticed in those experiments as well as in the present ones. Compare the initial rate of secretion in the glands perfused with cadmium, Figure 1 with that of the glands perfused without cadmium, Figure 2.

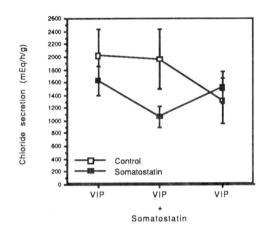

Figure 2. Somatostatin inhibition of VIP-stimulated chloride secretion. Somatostatin inhibited chloride secretion by 30% an amount not different from that observed in the presence of cadmium. Symbols as in Figure 1. Control, n=6, somatostatin, n=8. Values are mean ± SEM.

In the current experiments cadmium also inhibited the effect of VIP but again did not prevent the inhibitory effect of somatostatin. There is no explanation for the difference between these experiments and those reported by Forrest et al. It is clear from these results that cadmium, that interferes with the inhibitory and stimulatory effect of other peptides, does not interfere with the inhibitory effect of somatostatin. These results suggest that cadmium toxicity is exerted at many different levels.

Cadmium inhibits stimulated chloride secretion and reduces the inhibitory effect of NPY and the stimulatory effect of ANP. Since the effect of ANP is mediated by the release of VIP those findings suggested to us that cadmium interferes with neurotransmitter release and that in addition it has a direct inhibitory effect on the chloride secreting cell. It was important therefore to establish whether cadmium interfered with the inhibitory effect of peptides other than NPY to ascertain whether it blocks a common inhibitory pathway. The above results suggest that it does not. Further studies with other inhibitory peptides are necessary to establish this.

Supported by grants NIH DK 18078, and NIEHS ESO 3828 and the Pew, W.R. Hearst Foundation and Burroughs-Welcome Foundation grants to Heather Brignull, Sonya Hornung, and Douglas Wolff.

NEUROPEPTIDE INHIBITION OF ACTIVE TRANSPORT IN CULTURED RECTAL GLAND CELLS OF <u>SQUALUS</u> <u>ACANTHIAS</u>

F.H. Epstein[1], K. Spokes[1], P. Silva[2], R. Solomon[2], S. Hornung,
J.D. Stidham, A.S. McCraw, D.S. Nelson, K. Karnaky, Jr.[3]

Beth Israel[1] and Deaconess[2] Hospitals, Joslin Diabetes Center
and Harvard Medical School[1,2], Boston, MA 02215 and
Medical University of South Carolina[3], Charleston, SC 29425

The rectal gland of <u>Squalus</u> <u>acanthias</u> is innervated by nerves immunostaining for peptide inhibitors of secretion, including neuropeptide Y (NPY), bombesin and somatostatin. Somatostatin is thought to have a direct inhibitory effect on rectal gland cells, while bombesin's inhibitory action is at least partly indirect, since it releases somatostatin from the isolated perfused gland. Whether NPY inhibition is direct or indirect has been uncertain. We therefore tested the inhibitory actions of NPY, bombesin and somatostatin in primary monolayer cultures of shark rectal gland, in which nerve cells are thought to be absent, by measuring short-circuit current (I_{sc}), a direct index of chloride secretion in this preparation.

Monolayer cultures of rectal gland cells were prepared as previously described (Am J Physiol 260:C813-C823, 1991). Tubules prepared from minced rectal glands digested with collagenase were inoculated into 35 mm culture dishes containing type 1 collagen gels supported with nylon mesh (Small Parts, Inc., Miami, FL) 18 mm in diameter. Cultures were grown to confluence in approximately 10 days, mounted in Ussing chambers, bathed on apical and basal sides with shark Ringer's solution containing 5 mM glucose at pH 7.5 and studied using standard electrophysiological methods. Reagents were added to the basolateral side of the membrane in all experiments described.

Addition of forskolin 10^{-6}M to the basolateral surface of cultured rectal gland cells regularly increased I_{sc} by about 10 times, from 11.2 ± 7.6 uamp/cm^2 to 112.5 ± 25 (mean S.E.; n=6. p<0.001) without a consistent change in transepithelial resistance. NPY 10^{-6}M, applied after forskolin-induced stimulation had reached a plateau, consistently induced a fall in I_{sc} apparent within 2 minutes and maximum after 20 minutes, that averaged $29 \pm 4\%$ (n=6, p<0.01), without a significant change in membrane resistance. The inhibition produced by NPY was not reversed by 10^{-6}M glyburide, an inhibitor of the ATP-sensitive K$^+$ channel.

Somatostatin, 10^{-6}M, also inhibited I_{sc} in rectal gland culture that had previously been stimulated by forskolin. The percentage inhibition induced by somatostatin in I_{sc} averaged $52 \pm 10\%$ (n=4), with a parallel decrease in transmembrane voltage and no change in transepithelial resistance. By contrast, bombesin, 10^{-6}M, did not affect I_{sc} when applied to stimualted rectal gland cells in culture.

These results supply strong evidence that NPY inhibits rectal gland secretion of chloride by a direct action upon rectal gland cells, as does somatostatin. An inhibitory action of bombesin, on the other hand, cannot be demonstrated in cultured rectal gland cells, in accord with indirect evidence that bombesin's inhibition of whole perfused glands is secondary to the release of other neuroinhibitors including somatostatin.

Assisted by grants from the Burroughs-Wellcome Foundation (Stidham, McCraw, Nelson), the W.R. Hearst Foundation, and the Pew Trusts (Hornung) to MDIBL, from the NIH (DK18078) to F.H. Epstein, and from the Cystic Fibrosis Foundation to K. Karnaky, Jr.

C-TYPE ATRIAL NATRIURETIC PEPTIDES ARE POTENT DILATORS OF SHARK (SQUALUS ACANTHIAS) VASCULAR SMOOTH MUSCLE.

David H. Evans[1], Tes Toop[1], John Donald[1], John N. Forrest, Jr.[2]

[1]Department of Zoology, University of Florida. Gainesville, FL 32611
[2]Departmentof Medicine, Yale University School of Medicine, New Haven, CT 06510

A significant component of the array of physiological actions of members of the atrial natriuretic peptide (ANP) family of hormones is dilation of vascular smooth muscle, both in vivo and in vitro (e.g., Winquist and Hintze. Pharmacol Ther 48: 417-426, 1990). We have recently shown that the endothelium-free, vascular smooth muscle (VSM) from the ventral aorta of the dogfish shark, Squalus acanthias, is sensitive to synthetic, rat ANP (rANP; Evans, J Exp Biol 157: 551-555, 1991). The EC_{50} of the vasodilation is 7 nM, in the same range as that of ANP on isolated, mammalian VSM rings (e.g.,Winquist and Hintze, Op. Cit., 1990). Despite this rather high sensitivity of shark VSM to rANP, one might suggest that fish peptides might be even more effective. In fact, Takei et al. (Biochem. Biophys. Res. Comm. 164: 537-543, 1989; Biochem. Biophys. Res. Comm. 170: 883-891, 1990) have shown that both eel ANP and eel CNP (C-type natriuretic peptide) are 100 times as potent as rANP in production of hypotension in the eel in vivo. To date, Takei's studies on the eel are the only published data on the physiological effects of an homologous ANP-like peptide on a fish species. Recently, Schofield et al. (Am J Physiol 261: F734-F739, 1991) have described CNP from the heart of the spiny dogfish, Squalus acanthias which has 91% homology with CNP isolated from the heart of the European dogfish, Scyliorhinus canicula, (Suzuki et al. FEBS Lett 282: 321-325, 1991), and 82% homology with either porcine CNP (Sudoh et al., Biochem. Biophys. Res. Commun. 168: 863-870, 1990) or killifish CNP (Price et al., Biol. Bull. 178: 279-285, 1990). This recent sequencings of shark CNP's are especially interesting because they are the first to show that a CNP is present in cardiac tissue and, therefore, is presumably released and circulates as a hormone. The present study was undertaken to investigate the putative role of the homologous S. acanthias CNP in the control of gill hemodynamics in that species by examining the effect of the peptide on VSM rings from the ventral aorta. In order to test the specificity of the response, we also examined the efficacy of both killifish and porcine CNP on this system.

Endothelium-free, VSM rings were prepared and mounted in 10 ml of shark Ringer's solution as previously described (Evans and Weingarten, Toxicology 61: 275-281, 1990) The initial tension was approximately 500 mg. Rings were not preconstricted before the cumulative addition of shark, killifish or porcine CNP[1] to produce a concentration range of approximately 10 pM to 100 nM (depending on the agonist). Spiny dogfish shark CNP (sCNP) was diluted in phosphate buffered saline and added in small volumes directly to the experimental baths. Killifish CNP (kCNP; in HPLC solvent; 30% acetonitrile, 70% water, 0.1% trifluoroacetic acid) was aliquoted into polyethylene microfuge tubes, lyophilized in a Speedvac (Savant, Farmingdale, NY), and stored at -70 °C until used. Porcine CNP (pCNP; Peninsula Laboratories) was dissolved in 0.2 M acetic acid, aliquoted, lyophilized, and stored at -70 °C. Concentrated solutions were made up in small quantities of distilled water before addition to the shark Ringer's to produce the specific concentration. Maximal decreases in tension produced were: 16 ± 4.6 % (4) (mean ± S.E., (Number of rings)), 8.1 ± 1.2% (9), and 15 ± 4.6% (7) for sCNP, kCNP, and pCNP, respectively, all somewhat below the 24% recorded in an earlier study using rANP (Evans, Op. Cit., 1991). Curve fitting and apparent EC_{50} values were estimated using Cricket Graph (1.3; Cricket Software, Inc., Malvern, PA) and SuperPaint (2.0; Silicon Beach Software, San Diego, CA) on a Macintosh II microcomputer.

---

[1] Shark and killifish CNPs were synthesized by the Protein Cores at Yale and UF, respectively; procine CNP was purchased from Peninsula Labs, Belmont, CA.

Shark CNP produced a concentration-dependent reduction in tension in the isolated VSM from the shark ventral aorta, with an apparent EC_{50} of 0.5 nM. Thus, this tissue is nearly 15 times more sensitive to homologous CNP than it is to heterologous rANP (Evans, Op. Cit., 1991). The reduction in tension in the vascular rings produced by either kCNP or pCNP had approximately the same EC_{50}, suggesting that CNPs in general are more effective than ANPs in facilitating vasodilation in this species. This extreme sensitivity to CNPs by shark aortic VSM is surprising in light of the fact that pCNP is approximately 1/100th as effective as rANP in producing hypotension in intact rats (Sudoh et al., Op. Cit., 1990). In fact, recent radioimmunoassay data indicate that CNP is present only in the brain of the pig and human, not in the heart, consistent with the proposition that, in mammals, CNP is a neuropeptide, rather than a circulating hormone (Minamino et al., Biochem Biophys Res Commun 179: 535-542, 1991).

The first two fish CNPs (killifish and eel, both teleosts) also were isolated from brain tissue, but both the shark CNPs were isolated (the gene cloned in the case of S. acanthias; Schofield et al., Op. Cit., 1991) from cardiac tissue, suggesting that this might be a significant site of CNP synthesis in the Chondrichthyes. Furthermore, our immunohistochemical studies have found that antibodies which recognize rat ANP and porcine BNP show no immunoreactivity in sections of S. acanthias heart. In contrast, specific immunoreactivity is observed using an antibody that has cross-reactivity with all the known structural forms of natriuretic peptides, and this immunoreactivity is abolished by incubation of the antisera with pCNP (Donald, Vomachka, and Evans, this volume). Thus, recent data support the conclusion that, at least in the Chondrichthyes, CNP is produced in the heart and can produce relaxation of vascular smooth muscle, suggesting that it is a circulating hormone. It is unclear whether CNP is the only or even the dominant natriuretic peptide in this group, since earlier data demonstrating ANP-like immunoreactivity in the heart (Uemura et al., Cell Tiss. Res. 260: 235-247, 1990) and plasma (Evans et al., Am.J. Physiol., 257, R939-R945, 1989) of sharks did not differentiate between ANP and CNP.

The roles of CNP in shark hemodynamics and osmoregulation remain to be determined, but recent data suggest that it is 50-100 times more potent than rANP in the stimulation of Cl$^-$ transport by the S. acanthias rectal gland (Karnaky and Forrest, this volume), again in contrast to the relatively low natriuretic ability of pCNP in the rat (1/100th that of ANP; Sudoh et al., Op. Cit., 1990). Finally, if shark branchial vessels also vasodilate in response to CNP, as is apparently the case in teleosts (e.g., Evans et al., Op. Cit., 1989), the ultimate effect of the putative release of CNP from the heart would be an increase in gill perfusion. Since the branchial epithelium is the site of diffusional gain of NaCl and osmotic gain of water from the surrounding sea water (e.g., Evans, In: Comparative Physiology of Osmoregulation in Animals, edited by G. M. O. Maloiy. Orlando: Academic Press, 1979, vol. 1, p. 305-370.), the response to CNP release would be hypervolemia and hypernatremia, directly opposite to the natriuretic and hypovolemic response to the natriuretic peptides in mammals (e.g., Brenner, et al., Physiol. Revs. 70: 665-699, 1990.). Such an apparently osmotically-inappropriate effect of CNP in sharks suggests that this peptide hormone may have evolved in this group to control other physiological systems, such as gas-exchanger hemodynamics. Our findings support the conclusion that CNP is a hormone of some hemodynamic importance in sharks, despite its apparent exclusively-neurohumoral role in mammals.

These studies were supported in part by NSF DCB 8916413 (DHE), NIH DK34208 (JNF), and NIH EHS-P30-ESO3828 to the Center for Membrane Toxicity Studies.

IMMUNOHISTOCHEMICAL LOCALISATION OF NATRIURETIC PEPTIDES[*] IN TISSUES OF TELEOST, ELASMOBRANCH, AND CYCLOSTOME FISH

John A. Donald[1], Archie J. Vomachka[2] and David H. Evans[1]
[1]Department of Zoology, University of Florida, Gainesville, FL 32611
[2]Department of Biology, Beaver College, Glenside, PA 19038

The distribution of natriuretic peptide immunoreactivity was determined in the heart and brain of dogfish, Squalus acanthias, hagfish, Myxine glutinosa, and in the heart only of skate, Raja erinacea, flounder, Pseudopleuronectes americana and sculpin, Myoxocephalus octodecimspinosus. Heart tissues were fixed in 15% saturated picric acid and 4% formaldehyde in phosphate-buffered saline (PBS) at pH 7.4 for 16-24 h at 4 °C. Following fixation, tissues were washed in 80% ethanol to remove excess picric acid and dehydrated through an alcohol series, incubated in xylene, and rehydrated to PBS. Tissues were stored in PBS containing 20% sucrose, 3% polyethylene glycol (MW 400) and 0.1% sodium azide. Brains were removed and immersion-fixed in ice-cold 4% paraformaldehyde in 0.1M phosphate buffer at pH 7.4 for 1 h and then postfixed overnight in the same fixative. The brains were washed for 24 h in PBS containing 15% sucrose, and stored as above. Frozen sections were cut on a cryostat. Immunohistochemical staining was performed with the avidin-biotin-peroxidase complex (ABC) using an ABC kit (Vector Labs, Burlingame, CA, USA). Three antibodies were used: one raised against porcine brain natriuretic peptide (pBNP) which cross-reacts with rat atrial (rANP) and porcine C-type (pCNP) natriuretic peptides (termed natriuretic peptide-like immunoreactivity, NP-LI); the second raised against pBNP which cross reacts with CNP, but not ANP (termed pBNP-immunoreactivity, pBNP-LI); the third raised against rANP (termed rANP-like immunoreactivity, rANP-LI). The specificity of the immunohistochemical reactions was determined by incubating each antisera with either rANP (Bachem, CA, USA), pBNP (Peninsula, CA, USA), or pCNP (Peninsula, CA, USA). The antisera were incubated with 7-10 μg of each peptide at the working dilution for 24 h at 4 °C.

NP-LI and pBNP-LI, but not rANP-LI were observed in the heart of flounder and sculpin. In the shark and skate heart, NP-LI was observed in all cardiocytes in the atrium, but, only in a few cardiocytes in the ventricle, adjacent to the epicardium. No rANP-LI or pBNP-LI was observed in any cardiocytes of the shark and skate heart, and no immunoreactivity of any type in the branchial and portal hearts of hagfish. An extensive distribution of NP-LI perikarya and fibres was found in the brain of shark and hagfish (Table 1 for anatomical locations). Furthermore, in both species pBNP-LI perikarya and fibres were present in many areas which showed NP-LI but in a lower density. (Table 1). No rANP-LI immunoreactivity was found in the brain of shark and hagfish.

The observation of pBNP-LI in the heart of flounder and sculpin indicates that a pBNP/pCNP-like peptide is present in the heart, and could be released into the circulation to affect osmoregualtory tissues such as the gill, gut, and kidney. The more extensive distribution of NP-LI in the heart probably suggests the presence of an ANP-like peptide structurally different to the epitope recognised by the rANP antiserum. Since CNP has been demonstrated in the heart of Squalus (Schofield et al. 1991, Am. J. Physiol. 264: F734-F739), the presence of NP-LI in the heart shows that the antisera cross reacts with either native shark CNP or a natriuretic peptide different from rANP or pBNP. Although the chemical structure of natriuretic peptides in the brain of shark and hagfish are unknown, these observations show that a component of the natriuretic

---

[*] The term natriuretic peptide is used to describe the family of peptides which includes atrial natriuretic peptide (ANP), brain natriuretic peptide (BNP), and C-type natriuretic peptide (CNP)

peptide complement is similar to pBNP or pCNP. The presence of natriuretic peptides in the brain suggest they could be important neuromodulators and/or neurotransmitters. Furthermore, there appears to be divergence in the structural forms of natriuretic peptides in the heart and brain of shark and hagfish.

Table 1. Distribution of NP-LI and pBNP-LI in the brain of shark and hagfish*					
SHARK			**HAGFISH**		
Brain Region	NP-LI	pBNP-LI	**Brain Region**	NP-LI	pBNP-LI
Telencephalon			**Telencephalon**		
-olfactory bulbs	++	+	-olfactory bulbs	+	-
-subpallium	++	+	-pallium (layer 2)	+	±
-area superficialis basilis	+++	++	-prim. hippocampi	+++/p	+/p
-n. septi medialis	+++/p	++/p			
-area periventricularis	+++/p	-	**Diencephalon**		
-basal forebrain bundle	+++	+	-pars ventralis th.	++/p	+/p
-n. septi caudoventralis	+++	-	-n. diffusus hypoth.	++/p	+/p
-n. septi caudodosrsalis	+++	-	-pars dorsalis th.	++/p	±/p
			-hypoth.	+	±
Diencephalon			-n. tuberculi post.	++/p	+
-preoptic area	+++/p	++/p	-n. profundus	++/p	+
-ventral/dorsal th.	++/p	+/p			
-habenula	+	±	**Mesencephalon**		
-rostral hypoth.	++	+	-tectum	++	+
-optic nerves	-	-	-tegmentum	+	±
-neurointermediate lobe	++				
-adenohypophysis	-	-	**Rhombencephalon**		
			-n. vent tegementi	++/p	+/p
Mesencephalon			-lateral fibre tracts	++	+
-tectum mesencephali.	++	±			
-tegmentum	+	±	**Spinal cord**		
			-ventral/dorsal tracts	+	±
Cerebellum	-	-			
Rhombencephalon					
-griseum centrale	++/p	±/p			
-reticular formation	++	±			
Spinal cord					
-cornu ventrale	+	±			
-funiculus dorsalis	+	±			

hypoth. hypothalamus; nucleus; post. posteriosis; prim. primordium; th. thalamus.
* - no immunoreactive structures; ±, sparse; +, sparse to moderate; ++, moderately dense; +++, highly dense; p, perikarya present.

(Supported by NSF DCB 8916413 to DHE; an ROA to AJV; NIEHS-P30-ESO3828 to the Centre for Membrane Toxicity Studies)

C-TYPE NATRIURETIC PEPTIDE IS A POTENT SECRETAGOGUE FOR THE CULTURED SHARK (SQUALUS ACANTHIAS) RECTAL GLAND

Karl J. Karnaky, Jr.[1], James D. Stidham[2], David S. Nelson[2],
Andrew S. McCraw[2], John D. Valentich[3], Mike P. Kennedy[1,5],
and Mark G. Currie[4]

[1]Department of Anatomy and Cell Biology and the Marine Biomedical
and Environmental Sciences Program, Medical University of South
Carolina, Charleston, SC 29425

[2]Department of Biology, Presbyterian College, Clinton, SC 29325

[3]Department of Internal Medicine, University of Texas Medical
Branch, Galveston, TX, 77550

[4]Monsanto Company, St. Louis, MO 63167

[5]Grice Marine Biological Laboratory, Charleston, SC 29412

Elasmobranchs utilize the rectal gland to regulate plasma ion concentrations and fluid volume (Solomon et al., Am. J. Physiol. 248:R638-R640, 1985). In one model, atrial natriuretic peptide, released from the heart of the shark, causes the release of vasoactive intestinal peptide from peritubular nerve endings which results in the stimulation of chloride secretion (Silva et al., Am. J. Physiol. 252:F99-F103, 1987). We reported that atriopeptins (AP's) act directly on cultured shark rectal gland cells from apical or basolateral sides to stimulate sodium chloride secretion and elevate the second messenger, cGMP (Karnaky et al., Am. J. Physiol. 260:C1125-C1130, 1991).

Recently, C-type natriuretic peptides (CNP's) have been demonstrated in the heart of Scyliorhinus canicula (Suzuki et al., FEBS 282:321-325, 1991) and Squalus acanthias (Schofield et al., Am. J. Physiol. 261:F734-739, 1991). Since these CNP's differ by only several amino acids from killifish brain CNP (KCNP: Price et al., Biol. Bull. 178:279-285, 1990) we have tested the effects of the latter (a kind gift of Dr. D. H. Evans) on the cultured shark rectal gland.

Monolayer cultures of spiny dogfish (Squalus acanthias) shark rectal gland epithelium in 6-well culture plates were equilibrated with a low bicarbonate, HEPES-buffered Ringer solution containing 1 mM 3-isobutyl-1-methyl xanthine (IBMX) for 20 min. Cells were incubated for 10 min with varying concentrations of KCNP. Monolayer cultures of dogfish shark rectal gland epithelium maintained on collagen-coated nylon mesh were used for measuring short-circuit current (I_{sc}) in Ussing chambers (Valentich, Bull. Mt. Des. Isl. Biol. Lab., 26:91-94, 1986).

Basolateral exposure to 10^{-10} M KCNP markedly stimulated bumetanide-inhibitable I_{sc} [from a control value of 10.0 ± 2.0 to 40.0 ± 5.0 $\mu amp/cm^2$; n=6]. Higher doses did not increase I_{sc} appreciably. KCNP was also effective from the apical side. By comparison, 10^{-9} M rat AP III causes a only few $\mu amp/cm^2$ increase. In 4 experiments, 10^{-10} M KCNP addition resulted in elevated cGMP concentrations (from a control value of 5.0 ± 0.5 to a value of 12.0 ± 1.0 pmoles cGMP/mg protein). By comparison, 10^{-9} M rat AP III has no effect on cGMP in the cultured shark rectal gland. KCNP did not stimulate chloride secretion in the killifish operculum (N=6). In summary, KCNP is approximately 100 times more potent than rat AP III in stimulating chloride secretion and in elevating cGMP in the cultured SRG. The potency suggests that a C-type natriuretic peptide plays a role in controlling chloride secretion in the shark rectal gland.

Supported by a Cystic Fibrosis Foundation grant to K. Karnaky. J. Stidham, David Nelson, Andrew McCraw were Burroughs-Wellcome Fellows. Contribution No. 100 of the Grice Marine Biological Laboratory.

THE EFFECT OF NATRIURETIC PEPTIDES ON NA$^+$ AND CA^{2+} CHANNELS OF MAMMALIAN CARDIAC MYOCYTES

Lisa Ann Sorbera, Timothy Anderson, and Martin Morad
University of Pennsylvania, Department of Physiology,
Philadelphia, PA 19104-6085

Recently, we reported that ANP (rat, 1-28 a.a.) modulates cardiac Na$^+$ channels by making them more selective to Ca^{2+} (Sorbera and Morad, Science 247:969, 1990). Two other natriuretic peptides (BNP and CNP) with sequence homology to ANP have been identified and are secreted from the mammalian ventricle (Sudoh et al. Nature 332:78, 1988; Sudoh et al. Biochem. Biophys. Res. Commun. 168:863, 1990). We examined the effect of BNP-45 (rat, 51-95 a.a) and CNP (porcine, 22 a.a.) on I_{Na} and I_{Ca}. Single rat and guinea pig ventricular myocytes were enzymatically isolated (Mitra and Morad, Am. J. Physiol. 249: H1056-H1060, 1985) and whole cell clamped (Hamill et al. Pflügers Arch. 391:85-100, 1981). Ventricular cells with capacitances ranging from 40 to 140 pF were used. The standard external solution contained (in mM): 137 NaCl; 1 MgCl$_2$; 2 CaCl$_2$; 10 HEPES; 10 glucose titrated to pH 7.4 with NaOH. In order to control I_{Na}, [Na$^+$]$_o$ was reduced to 10 mM and replaced with 127 CsCl. Patch pipettes with resistances of 1 to 3 MΩ were used. The standard internal dialysate contained (in mM): 10 NaCl; 110 CsCl; 5 MgATP; 20 HEPES; 14 EGTA titrated to pH 7.2 with CsOH. In internal solution where GTP (guanosine triphosphate) was added, 1.0 mM MgCl$_2$ in addition to 5 mM MgATP was also included. A rapid (< 50 ms), electronically controlled concentration clamped system was used to externally deliver the hormones to the individual myocyte. I_{Na} was activated by 15 to 30 ms depolarizing pulses applied at 5 s intervals of -30 or -40 mV from holding potentials of -70, -80 or -90 mV. In Figure 1A, I_{Na} was activated by a depolarizing pulse to -30 mV from a holding potential of -90 mV. BNP (10 nM), like ANP, suppressed I_{Na} rapidly (< 50 ms) and reversibly 40.7 ± 2.5 % (n=44) in a dose-dependent manner and at all voltages tested. BNP also shifted the steady state inactivation of I_{Na} by 10-15 mV (Figure 1C), but unlike ANP, BNP did not alter the selectivity of the channel as measured by the change in the reversal potential of I_{Na} (Figure 1B).

During continuous exposure to BNP, I_{Na} slowly recovered, suggesting possible desensitization of the response. CNP (1-100 nM) had no consistent effect on I_{Na} in the same set of cells. In GTP (0.5 mM) dialyzed cells, BNP's suppressive effect on I_{Na} was reduced to 10.1 ± 1.8 % (n=15). Dialysis with 1 mM GDPß-S (guanosine 5'-O-(2-thiodiphosphate) did not alter the suppressive effect on I_{Na}.

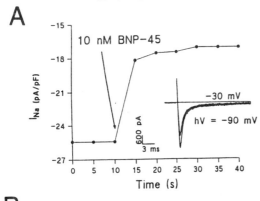

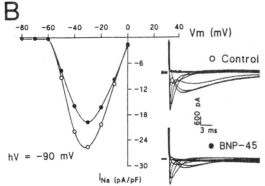

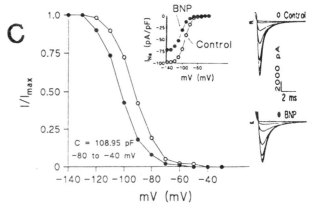

We also examined the effect of BNP and CNP on I_{Ca}. I_{Ca} was activated at 5 s intervals by 15 to 30 ms depolarizing pulses to 0 mV from holding potentials of -30, -40, -50, or -60 mV. In Figure 2A, I_{Ca} was activated by a depolarizing pulse to 0 from a holding potential of -60 mV. BNP (10 nM) rapidly enhanced I_{Ca} 163.40 ± 7.1 % in some cells (n=37, Figure 2B), but in others suppressed I_{Ca} by 21.44 ± 1.5 % (n=38). In cells dialyzed with 0.5 mM GTP, BNP only suppressed I_{Ca} by 25.27 ± 11.4 % (n=7); and no enhancing effect was ever found. BNP had little or no effect on I_{Ca} in cells dialyzed with GDPß-S. CNP, once again, had no consistent effect on I_{Ca}.

Thus when the Na^+ or Ca^{2+} channels were phosphorylated, the effect of BNP on the Na^+ channel was significantly reduced or completely reversed in the case of the Ca^{2+} channel. Our experiments suggest specific and differential effects of natriuretic peptides ANP, BNP, and CNP on the Na^+ and Ca^{2+} channel. While ANP regulates the Na^+ channel by altering its selectivity, BNP modulates the Na^+ channel by regulating its gating. CNP, on the other hand, appears to have minimal involvement in the regulation of cardiac Ca^{2+} and Na^+ channels.

Supported by NIH grant HL16152 and a grant from the AHA Maine affiliate. Technical support of Brooke Maylie is acknowledged.

A

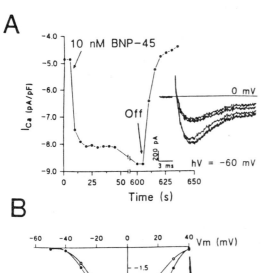

B

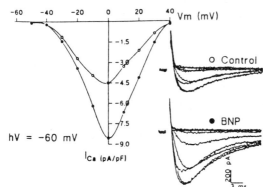

AN ATRIAL NATRIURETIC PEPTIDE-LIKE FACTOR ISOLATED FROM THE DOGFISH SHARK (SQUALUS ACANTHIAS) RECTAL GLAND: INITIAL CHARACTERIZATION

Karl J. Karnaky, Jr.[1], William F. Oehlenschlager[2], James D. Stidham[3], David S. Nelson[3], Andrew S. McCraw[3], John D. Valentich[4], Mike P. Kennedy[1,5], and Mark G. Currie[6]

[1]Department of Anatomy and Cell Biology and the Marine Biomedical and Environmental Sciences Program and [2]the Department of Pharmacology, Medical University of South Carolina, Charleston, SC 29425

[3]Department of Biology, Presbyterian College, Clinton, SC 29325
[4]Department of Internal Medicine, University of Texas Medical Branch, Galveston, TX, 77550
[5]Grice Marine Biological Laboratory, Charleston, SC 29412
[6]Monsanto Company, St. Louis, MO 63167

The elasmobranch rectal gland helps regulate plasma ion concentrations and fluid volume (Solomon et al., Am. J. Physiol. 248:R638-R640, 1985). We have reported that atriopeptins (AP's) act _directly_ on cultured shark rectal gland cells from apical or basolateral sides to stimulate chloride secretion and elevate the second messenger, cGMP (Karnaky et al., Am. J. Physiol. 260:C1125-C1130, 1991). More recently, we have shown that C-type natriuretic peptide is a potent stimulator of chloride secretion when added to either apical or basolateral side (Karnaky et al., this Bull., 1992). C-type natriuretic peptides have been demonstrated in the heart of Scyliorhinus canicula (Suzuki et al., FEBS 282:321-325, 1991) and Squalus acanthias (Schofield et al., Am. J. Physiol. 261:F734-739, 1991).

The ability to regulate cell function from the apical side of secretory cells in the shark rectal gland suggests novel new control mechanisms which involve local secretion of autocrine or paracrine factors. These findings raise interesting questions concerning the origin and nature of peptide hormones which carry out this function. We have prepared extracts of the shark rectal gland and tested these with three different bioassays.

Extracts were made in the following manner: shark rectal glands (a kind gift of Dr. Michael Field) were minced, boiled, acidified, homogenized, and centrifuged. The supernatant was then applied to a C18 Sep-Pak cartridge. Two partially-purified fractions were obtained by successive rinses in 10% and 40% acetonitrile, producing SP10 and SP40 fractions, respectively. Our preliminary data has shown that the SP40 fraction has proven the best fraction for isolating natriuretic factors from dogfish shark heart and we and others have routinely used this type of protocol to partially purify atrial natriuretic peptide (ANP) and ANP-like peptides from hearts and plasma of many different species.

In a competition binding assay we utilized bovine pulmonary endothelial cells (BPAE cells), which are rich in ANP-binding sites, to examine the competition between our extract and ^{125}I-ANP for binding. ^{125}I-ANP (^{125}I-rAPIII) binds to these cells in a highly specific manner (specific binding $\geq$ 95% of total binding) with a K_D=0.2 nM. Unrelated peptides such as insulin, bradykinin, angiotensin II, and dynorphin do not displace ^{125}I-ANP from these sites at concentrations as high as 1 μM. These cultures were treated with ^{125}I-ANP only or with this isotope and varying concentrations of the SP10 or SP40 fractions. Non-diluted SP40 fractions displaced 83% of the control ^{125}I-ANP binding (2590 ± 180 c.p.m. bound with the SP40 present versus a control value of 15,260 ± 690 c.p.m. bound; N= average of two samples). A dose-response was demonstrated as serially-diluted SP40 fractions were correspondingly less effective at competing for the ANP-binding sites on the endothelial cells. The resultant displacement curve roughly paralleled that obtained using unlabeled synthetic ANP. As expected the SP10 fraction was less effective in displacing ^{125}I-ANP from its binding sites with only the undiluted SP10 fraction demonstrating activity (23% displacement). An equivalent displacement was obtained with a 16-fold dilution of the SP40 fraction.

C18 Sep-Pak purified extracts were also tested for their effects on intracellular cGMP levels in shark rectal gland (SRG) cells. ANP is known to increase cGMP in these cells. The SP40 fraction increased intracellular cGMP levels over 10 fold (from a control value of 66.0 ± 8.2 to 794.0 ± 23.7; N=3), whereas the SP10 fraction had minimal effect. In our experience with shark heart, this SP10 fraction does not contain ANP-like molecules. Moreover, in a parallel purification protocol, an equivalent amount of the original acid extract was precipitated with 75% acetone and the supernatant from this fraction was found to elevate cGMP over 6 fold (from 66.0 ± 8.2 to 418.0 ± 54.0; N=2). In our experience, ANP-like peptides are typically found within this supernatant under these conditions. These results are consistent with the presence of an ANP-like peptide in SRG extracts; however, the possibility that an unrelated molecule is responsible for this increase in cGMP cannot be excluded.

Lastly, we tested the SP40 extract for possible stimulation of chloride secretion (as measured by the short-circuit current [Isc]) in cultured shark rectal gland cells. Eight shark rectal gland cultures showed significant increases in short-circuit current when the extract was added to the **APICAL** side of the cultures (Isc increased 8.8 ± 2.3 μamp/cm^2, P < 0.005, N=9). VIP and 2-chloro adenosine were not active from the apical side but were very potent secretagogues from the basolateral side in these cultures (VIP: 36.0 ± 6.5, P < 0.01, N=5; 2-chloroadenosine: 24.6 ± 3.5, p < 0.005, N=5).

In summary these data, taken together, strongly suggest that the SRG contains an ANP-like molecule which can control chloride secretion from the apical side of the epithelium. At present, it is unclear whether this ANP-like activity represents ANP-like molecule(s) synthesized within the SRG, synthesized at distant sites, e.g. heart, and sequestered in the SRG, or blood contamination. Regardless of tissue origin, the observations that the SRG contains extractable ANP-like activity and that these extracts are capable of stimulating Isc to the apical surface of SRG cells suggests that unique mechanisms for regulating Cl$^-$

secretion may be at work in the dogfish SRG. It has only recently been appreciated that several hormones work from not only the basal (blood) side of the epithelium, but also from the apical (lumenal) side as well. The ability to regulate cell function from the apical side suggests novel new control mechanisms which involve local secretion of autocrine or paracrine factors. Apical membrane hormone receptors have been described in a number of epithelia: 1) adenosine in the colonic epithelial cell line T_{84} (Barrett et al., Am. J. Physiol. 256:C197-C203, 1989); 2) lysylbradykinin in cultured rat epididymal epithelium (Cuthbert and Wong, J. Physiol. (Lond). 378:335-345, 1986; 3) insulin and insulin-like growth factor in colon (Pillion et al., Am. J. Physiol. 257:E27-E34, 1989; 4) vasoactive intestinal peptide in bovine tracheal epithelium (Elgavish et al., Life Sci. 44:1037-1042, 1989); 5) ANP in SRG epithelium (Karnaky et al., Bull. Mt. Des. Isl. Biol. Lab. 29:86-87, 1990); 6) bradykinin in canine trachea (Leikauf et al., Am. J. Physiol. 248:F48-F55, 1985). Two interesting recent papers suggest an autocrine or paracrine role for ANP in the kidney. McKenzie et al. (Amer. J. Anat. 190:182-191,1991) have used immunocytochemical techniques and revealed ANP within intercalated cells of the outer medullary and cortical collecting tubules and ducts of several adult mammals, including humans. The authors suggested that the intracellular localization of ANP may be the result of endogenous synthesis and that following secretion, ANP may be available to receptors in the inner medullary collecting ducts. Ritter et al. (J. Clin. Invest. 87:208-212, 1991) have detected constitutive secretion of an atriopeptin-like prohormone in the cortical tubule fraction in primary adult rat kidney cultures. These authors hypothesize that the renal ANP may be important as an autocrine or paracrine regulator of renal function. The cultured SRG epithelium should prove to be an excellent model system with which to understand these novel mechanisms.

Supported by a grant from the Cystic Fibrosis Foundation to K. Karnaky. J. Stidham, David Nelson, Andrew McCraw were Burroughs-Wellcome Fellows. Contribution No. 101 of the Grice Marine Biological Laboratory. We thank Kyle Suggs for excellent technical assistance.

CADMIUM AUGMENTS THE STIMULATION OF CHLORIDE SECRETION BY FORSKOLIN IN THE PERFUSED RECTAL GLAND OF SQUALUS ACANTHIAS

J.N. Forrest, Jr., D. Opdyke, C. Aller and G.G. Kelley
Department of Medicine, Yale University School of Medicine
New Haven, CT 06510

We previously reported that cadmium reversibly blocks the effects of multiple agonists that inhibit forskolin-stimulated secretion in the rectal gland, including adenosine (Forrest et al., Center for Membrane Toxicology Studies Annual Report 1986), peptide YY (Grasso et al. Bull MDIBL 29:57, 1990) and somatostatin (Forrest et al. Bull MDIBL 30: 117, 1991). Blocking of the inhibitory effects of somatostatin was present even at low concentrations (5-25μM) of cadmium. We noted that the effects of cadmium to block the inhibitory effects of these agonists (adenosine, peptide YY, and somatostatin) in each instance occurred after a delay of 8-10 min following the addition of forskolin and the agonist (see Forrest et al. Bull MDIBL, 57, 1991) One interpretation of these studies is that cadmium blocks inhibition, at least in part, by augmenting forskolin-mediated secretion.

In the present studies, we examined whether low concentrations of cadmium could augment basal or forskolin-stimulated chloride secretion in the isolated perfused rectal gland.

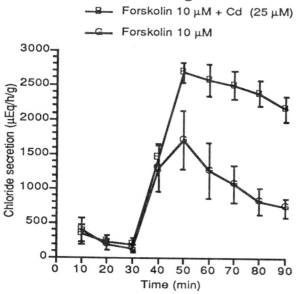

In all experiments cadmium was added to the perfusion medium at the onset of the experiment. As shown in Figure 1 (left) cadmium (25μM) had no detectable effect on basal chloride secretion in the perfused gland. When forskolin (10 μM) was added after 30 min of basal secretion, the presence of cadmium (25μM) resulted in a definite and sustained augmentation of forskolin-stimulated chloride secretion (see Figure 1) This augmentation of secretion was not present during the first 10 min of perfusion with forskolin but was evident at all subsequent time points examined (50-90 min).

Figure 1. Effects of cadmium on basal and forskolin-stimulated chloride secretion (p>0.05 at 40 min and p<0.01 at 60-90 min).

These findings suggest that at least a component of the effects of cadmium to block multiple inhibitory agonists (adenosine, peptide YY, somatostatin) may be explained by an effect of cadmium to augment forskolin-stimulated secretion. The diterpene forskolin stimulates adenylate cyclase by directly activating the catalytic unit of cyclase and by interacting with the coupling of the Gs alpha subunit with the catalytic subunit. Both GTP binding proteins and the catalytic subunit of cyclase are known to have sites that are sensitive to divalent cations. Thus, the observed response could be due to an effect of cadmium on G protein or catalytic unit function or an effect of cadmium to block either the release or action of local inhibitory autacoids.

This work was supported in part by NIH EHS-P30-ES03828 (Membrane Toxicity Studies) and NIH DK 34208 (Dr Forrest).

CHARACTERIZATION OF A LOW THRESHOLD CA^{2+} CHANNEL IN VENTRICULAR MYOCYTES FROM SQUALUS ACANTHIAS

James Maylie[1] and Martin Morad[2]

[1]Department of OB/GYN, Oregon Health Science University, Portland, OR 97201
[2]Department of Physiology, University of Pennsylvania, Philadelphia, PA 19104

Two types of Ca^{2+} channel currents, I_{Ca}, have been described in ventricular myocytes from Squalus acanthias (Mitra and Morad, MDIBL Bulletin 24:14-15, 1984; Dukes et al. MDIBL Bulletin 29:100, 1989). The differential sensitivity of the two I_{Ca} to holding potential and inorganic Ca^{2+} channel blockers in Squalus acanthias is similar to that reported in mammalian myocytes (Dukes et al. MDIBL Bulletin 29:100, 1989). No information, however, is available on the kinetics and hormonal regulation of I_{Ca} in cardiac myocytes from Squalus acanthias. In this report, we investigated the kinetics of inactivation of the T-type I_{Ca} and the hormonal modulation of both I_{Ca} in Squalus acanthias.

Single ventricular myocytes, isolated from Squalus acanthias (Mitra and Morad, Am. J. Physiol. 249:H1056-H1060, 1985) were studied with the whole cell patch clamp configuration. I_{Ca} was recorded under conditions that minimized ionic fluxes through Na$^+$ and K$^+$ channels. The patch pipette contained (mM): CsCl 240, MgCl$_2$ 1, Urea 300, EGTA 20, HEPES 20, TMAO 70, MgATP 5, pH 7.2. The standard external solution bathing the cells was (mM): NaCl 270, KCl 4, MgCl$_2$ 3, KH$_2$PO$_4$ 0.5, Na$_2$SO$_4$ 0.5, Urea 350, HEPES 10, glucose 10, CaCl$_2$ 3, pH 7.2. Following formation of whole cell recording configuration the external solution was switched to a Na$^+$ and K$^+$ free solution: TEA-Cl 275, CaCl$_2$ 5, MgCl$_2$ 5, HEPES 10, Urea 350, pH 7.2. Series resistance was compensated electronically (Axopatch 1D) and linear leak and capacitance currents were digitally subtracted with an ensemble of 16 hyperpolarizing pulses from the holding potential obtained before each test pulse.

The size of the calcium current activated during a test pulse depends on the prepulse potential preceding the test pulse. The dependence of the calcium current on the prepulse potential is referred to as the voltage dependence of inactivation or availability. The two types of I_{Ca} in Squalus were identified by their difference in voltage dependence of availability. Figure 1 shows I_{Ca} recorded at -40 mV (upper traces) and 20 mV (lower traces). Each test pulse was preceded by a 2 second prepulse to potentials ranging between -120 and -30 mV. A test pulse to -40 mV primarily activates a low threshold, T-type, I_{Ca}. Increasing the prepulse potential decreases the peak current during the test pulse to -40 mV monotonically. A single Boltzmann fit gives a voltage for half maximal inactivation of -69 mV and a steepness factor of 4 mV for e-fold change in peak current. The test pulse to 20 mV activates both T- and L-type I_{Ca}. Increasing the prepulse potential decreases I_{Ca} at 20 mV in two phases; a reduction in T-type between -90 and -50 mV and L-type between -40 and 0 mV. The continuous curve for the data points to 20 mV represents a fit of a sum of two Boltzmann distributions. The half maximal potential for inactivation was -67 mV for the T-type component and -22 mV for the L-type component at 20 mV. Inactivation of the T-type I_{Ca} is thus more negative and more steeply voltage dependent than the L-type I_{Ca}. The difference in voltage dependence of availability between T- and L-type I_{Ca} in Squalus is much larger than reported for mammalian and other vertebrate hearts.

Separation of the T- and L-type I_{Ca} is obtained by comparison of currents to a test pulse potential without (traces labelled h left panel of figure 2) and with a 2 sec prepulse to -65 mV (traces labelled p). The difference (h-p) between the two currents at each test pulse potential gives the T-type I_{Ca} (middle panel, figure 2). The current recorded with the prepulse (p) gives the L-type I_{Ca}. The peak current of the T-type (closed symbols) ant L-type (open symbols) is plotted in figure 2 (right panel). The threshold for activation of the T-type I_{Ca} is near -70 mV and that for the L-type I_{Ca} is near -40 mV. The current-voltage relation (IV) peaks near -40 mV for the T-type I_{Ca} compared to 0 mV for the L-type I_{Ca}. Normalization of the current by cell capacitance gives a peak T-type current density of 11 pA/pF which is approximately 10 times larger than that reported in mammalian atrial cells under similar conditions (Bean, J. Gen. Physiol. 86:1-30, 1985). The current density of L-type I_{Ca} of 11 pA/pF is comparable to that recorded in mammalian myocytes. The two types of I_{Ca} also are differentially sensitive to dihydropyrydines as in vertebrate cells; 1 μM nisoldipine blocks the L-type I_{Ca} but not the T-type I_{Ca}.

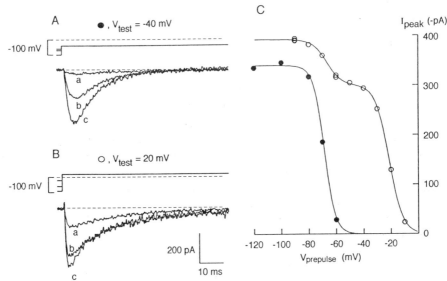

Figure 1.   Voltage dependence of availability of I_{Ca}. The membrane potential was held at different prepulse potentials for 2 sec and then clamped to a test potential of -40 mV (panel A) and 20 mV (panel B). A). Traces labeled a-c are current records for prepulse potentials of -100 (c), -70 (b), -60 (a) mV (voltage template above traces). B). Current records for prepulse potentials of -90 (c) ,-60 (b) ,-20 (a) mV. C). Plot of peak inward current (inverted) during test pulse versus prepulse potential. Closed symbols for test pulse to -40 mV and open symbols for test pulse to 20 mV. The continuous curve represents fit of the Boltzmann relation, $1/(1+\exp^{(V-V_{1/2})/k})$, to the data points. For test pulse to -40 mV, $V_{1/2}$ = -69 mV and steepness factor = 4.0 mV. For test pulse to 20 mV, a double Boltzmann was applied, $V_{1/2}$ =-67 mV, k=4.9 mV and $V_{1/2}$ =-22 mV, k=5.0 mV for the first and second component, respectively. Cell capacitance equal 44 pF.

The kinetics of inactivation of the T-type I_{Ca} is more rapid than the L-type I_{Ca} (compare prepulse traces to difference traces in figure 2) and more rapid than T-type I_{Ca} recorded in vertebrate myocytes. Time constants of inactivation (τ_{inact}) were determined from exponential fits to inactivation of the T- and L-type I_{Ca}. A plot of τ_{inact} of the L-type I_{Ca} versus voltage (figure 3) shows that the rate of inactivation is fastest near the peak of the IV relation (figure 2). As in vertebrate myocytes, the similarity between the voltage dependence of τ_{inact} and L-type I_{Ca} suggests that inactivation in part depends on current density. The values for τ_{inact} are similar to vertebrate myocytes, but a direct comparison is difficult because of the dependence of τ_{inact} on intracellular and extracellular conditions.

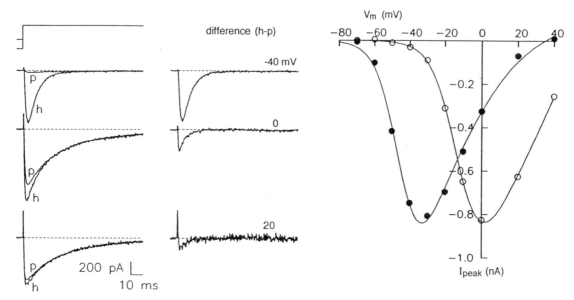

Figure 2.   Separation and current-voltage relation of T- and L-type I_{Ca}. Left panel: currents recorded at test potentials of -40, 0 and 20 mV from a holding potential of -120 mV (h) and following a 2 sec prepulse to -65 mV (p). Middle panel: difference between currents recorded without and with the prepulse (h-p). Right panel: plot of peak of the difference current h-p (closed symbols) and current with prepulse p (open symbols) versus test pulse potential. Cell capacitance was 74 pF, time constant of capacitive current was 0.09 ms and residual series resistance was 1.2 MΩ.

A plot of τ_{inact} of the T-type I_{Ca} versus voltage (figure 3) reveals that τ_{inact} reaches a voltage independent value of about 5 ms at positive test potentials. Similar results have been reported in dog atrial myocytes (15 ms, Bean, J. Gen. Physiol. 86:1-30, 1985) and fibroblasts (9 ms, Chen & Hess, J. Gen. Physiol. 96:603-630, 1990). The current interpretation is that the channel inactivation step is not voltage sensitive as originally suggested by Bezanilla and Armstrong (J. Gen. Physiol. 70:549-566, 1977) for Na$^+$ channels in squid axon. Kinetics of activation were also studied. Exponential fits to activation near the threshold of I_{Ca} required more than one exponential. This is consistent with recent studies suggesting that activation of T- and L-type I_{Ca} involves more than one closed state. The time constant of activation reaches a voltage independent value of about

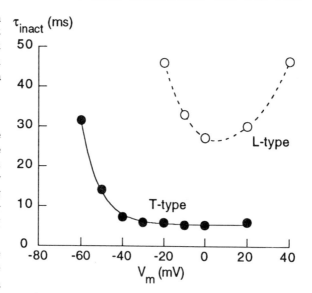

Figure 3. Voltage dependence of time constant of inactivation of T-type (closed symbols) and L-type (open symbols) I_{Ca}. From cell in figure 2.

0.4 ms at positive test potentials for both T- and L-type I_{Ca}. Thus at least one of the closed state transitions is not voltage sensitive as suggested for

T-type I_{Ca} in fibroblasts cells (Chen & Hess, J. Gen. Physiol. 96:603-630, 1990) which appears to be similar in T- and L-type channels.

These results show that the T-type I_{Ca} in <u>Squalus</u> activates at more negative potentials and is kinetically faster than T-type I_{Ca} in other tissues. In mammalian myocytes, a TTX-sensitive Na^+ channel, I_{Na}, is responsible for initiation of the action potential. The more negative threshold of the T-type I_{Ca} in <u>Squalus</u> suggests that its activation may contribute to initiation of the action potential. Currents were measured with Na^+ in the external solution. A rapidly activating inward current, sensitive to TTX, had a threshold near -45 mV. The T-type I_{Ca} had a threshold near -65 mV under this conditions. The T-type I_{Ca} is thus the first inward current activated by subthreshold depolarizations leading to initiation of the action potential <u>Squalus</u>.

Modulation of I_{Ca} by hormones is an important regulatory mechanism of heart rate and contractile force. The ß-agonist, isoproterenol (1 µM), increased L-type I_{Ca} 3-fold. The increase in L-type I_{Ca} by isoproterenol was reduced by 100 µM acetylcholine or carbachol. Neither isoproterenol or acetylcholine modulated the T-type I_{Ca}. Aterenol (10 µM) which has both α- and ß-adrenergic effects increased the L-type I_{Ca} by 30% and had no effect on the T-type I_{Ca}. These findings are similar to those observed in mammalian myocytes and show that the same regulatory pathways exist in both phyla.

These results show that the two types of I_{Ca} recorded in <u>Squalus</u> are in many respects similar to those recorded in other phyla. The modulation of L-type I_{Ca} and its voltage dependence are similar in mammals. The T-type I_{Ca} in <u>Squalus</u>, however, has properties that are quite different than T-type currents recorded in mammalian tissue. The main differences are; 1) the rate of inactivation is faster, 2) the threshold of activation and voltage dependence of availability is more negative, and 3) the peak current density is approximately 10 times larger than in other heart cells. Functionally, the T-type I_{Ca} will be important in initiation of the action potential and may influence excitation-secretion coupling in heart cells from <u>Squalus</u> which contain naturetic peptides.

This research was sponsored by a fellowship from the Lucille P. Markey Charitable Trust to JM and by a grant from AHA Maine to MM. The technical assistance of Brooke Maylie and Tim Anderson is gratefully acknowledged.

RECOVERY OF CALCIUM CHANNELS FOLLOWING RAPID PHOTOINACTIVATION OF DIHYDROPYRIDINES IN MAMMALIAN VENTRICULAR MYOCYTES

Luis Gandía, Lars Cleemann and Martin Morad
University of Pennsylvania, Department of Physiology
Philadelphia, PA 19104-6085

Dihydropyridine (DHP) calcium channels blockers are highly light sensitive. The aromatization of the dihydropyridine ring by UV light causes inactivation of the DHP and loss of their calcium channel blocking property. We have studied the kinetics of the recovery of calcium currents from nisoldipine block when a brief (180 μsec) light pulse was used to photoinactivate the drug. Inactivation of nisoldipine was independent on the intensity of the light pulse in the range of 50-200 J. In addition, the recovery of calcium current following photoinactivation of nisoldipine was dependent of the interval between the light pulse and the activation of the calcium current.

Single rat ventricular myocytes were enzymatically dissociated (Mitra and Morad, Am. J. Physiol. 249: H1056-H1060, 1985). Experiments were performed at room temperature (20-22 °C) using the whole cell patch-clamp technique (Hamill et al., Pflügers Arch. 391: 85-100, 1981). Patch pipettes had a resistance of 2-3 MΩ when filled with an intracellular solution containing (in mM): 10 NaCl, 100 CsAsp, 20 TEACl, 5 MgATP, 14 EGTA, 20 HEPES, 0.1 cAMP, pH 7.2. Extracellular solutions were exchanged rapidly (<50 msec) by using an electronically controlled concentration clamp system. Nisoldipine (1-10 μM) was added to the extracellular solution containing (in mM): 10NaCl, 127 CsCl, 1 MgCl$_2$, 2 CaCl$_2$, 10 HEPES, pH 7.4 and the solution protected from ambient light. Photolysis of DHP was achieved by the use of a Xenon arc lamp (240 J maximum output). Flashes (duration of about 0.180 msec) were focused onto the cell under investigation as described by Morad et al. (Mature 304, 635-638, 1983).

Figure 1 shows the effect of a light pulse of different intensity applied 10 msec prior to activation of Ca^{2+} channel by a 25 msec depolarizing pulse from -90 to 0 mV. Following addition of nisoldipine (10 μM) which blocked the Ca^{2+} current strongly, photoinactivation of the drug led to rapid recovery of I$_{Ca}$. The development of the block was significantly slower than unblocking the channel following the photoinactivation of nisoldipine. Maximum and often full recovery of Ca^{2+} current was consistently found only with the second depolarizing pulse following the light pulse. Figure 1 also shows significant difference in time course of nisoldipine-induced block of Ca^{2+} channel prior and following the

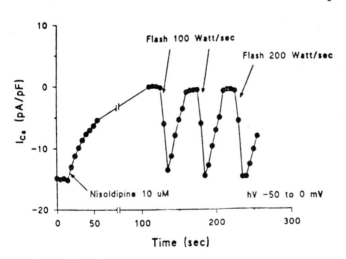

photoinactivation of nisoldipine. This discrepancy in the time course of the block may be in part related to equilibration of the membrane with lipophilic drug following the 1st exposure to the drug.

The time course of recovery of Ca^{2+} current within the time course of depolarizing pulse was also tested by applying the photoinactivating pulse prior to, during and following the depolarizing pulse. Figure 2 shows original traces recorded in presence of nisoldipine (N) and after the application of a light pulse of 100 J (arrow) at different times during the depolarizing pulse (F). Maximum recovery was always reached at the second depolarizing pulse (2) after the light pulse application.

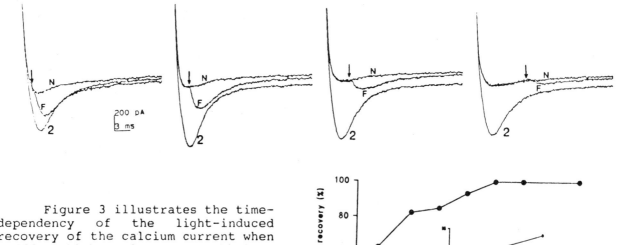

Figure 3 illustrates the time-dependency of the light-induced recovery of the calcium current when a 100 J light pulse was applied at different times before the application of a depolarizing pulse to 0 mV from a holding potential of -50 mV. Maximum recorevy was reached when the light pulse preceded 1.5-2 sec the depolarizing pulse. Similar results were obtained at a holding potential of -90 mV. Inset shows data for the first 400 msec using a different scale.

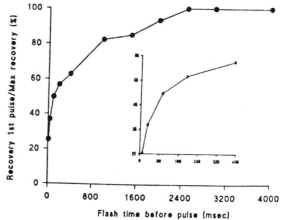

Our results showing that current recovery is bigger when photoinactivation of nisoldipine occurs earlier during the depolarizing pulse supports the idea that DHP calcium channel antagonists promote the movement of the Ca^{2+} channel to a drug-ocluded inactivated state (Morad and Rendt, Exp. Brain Res. (Suppl): 112-123, 1986). It has been suggested that DHPs block the Ca^2 channel by enhancing the rate of inactivation of the channel. This may explain why flashes given early during depolarization are more effective in eliciting large recovery than those applied later during the depolarization pulse. With subsequent repolarization channels return to the closed state and are then available for full activation upon depolarization.

Our results are consistent with a scheme where DHP type calcium channel blockers appear to block the calcium channel by driving it into a drug-occluded inactivated state.

Supported by NIH grant no. HL16152, AHA Maine affiliate to M.M. L. Gandía was supported by the Spanish Ministry of Education and Science. Technical support of Brooke Maylie and Tim Anderson are acknowledged.

TAURINE TRANSPORT AND RED CELL VOLUME IN HAGFISH (<u>MYXINE GLUTINOSA</u>)

Susan R. Brill[1], Mark W. Musch[2], Jennifer Alley[3], and Leon Goldstein[1]
[1]Department of Physiology, Brown University, Providence, RI 02912
[2]Department of Medicine, University of Chicago, Chicago, IL 60637
[3]Mount Desert Island Biological Laboratory, Salsbury Cove, ME 04672

The mechanism of volume-activated (VA) taurine release from hypotonically stressed cells has not been definitively established, although studies using inhibitors of the anion exchanger, band 3, have suggested that this protein is involved in the response. Since the RBC of hagfish are deficient in band 3 (Ellory et al. J. Exp. Biol. 129:377, 1987), we examined the effects of hypoosmotic exposure on taurine transport and cell volume in these RBC.

Blood was collected from skates and hagfish through a caudal vessel using a heparinized syringe. RBC were separated by centrifugation, washed with isosmotic incubation media, and resuspended in appropriate media for use in experiments. To determine the amount of band 3 in hagfish compared with skate RBC, cell membranes were prepared by high speed centrifugation and were incubated with ^{3}H$_2$DIDS, which binds with high affinity to band 3. After 30 min membranes were washed free of unbound ^{3}H$_2$DIDS, and the proteins separated by SDS-PAGE. Radioactivity in the various membrane proteins was quantified by liquid scintillation counting of digested gel slices. Skate RBC showed two peaks for bound ^{3}H$_2$DIDS. The major peak, representing 95% of the total radioactivity, was located on band 3. The remainder of the bound ^{3}H$_2$DIDS was found in a small peak, most likely representing glycophorin. In contrast, hagfish RBC membranes showed only one small peak of bound ^{3}H$_2$DIDS, containing less than 10% of the radioactivity found in skate RBC, and all localized to band 3. Additional evidence for low level of band 3 in hagfish RBC was found in sulfate transport studies. We used sulfate uptake to determine band 3 activity. We added 0.2ml of 20% RBC suspension to 2.5ml of either isosmotic or hypoosmotic incubation medium containing luCi ^{35}SO$_4$. The solutions were incubated at 15° C in a shaking bath and samples were collected at several timepoints. Hagfish RBC showed only 9% the amount of sulfate uptake of skate RBC at 60 min.

We next examined the effect of band 3 deficiency on VA taurine transport in hagfish RBC. When hagfish RBC are exposed to medium approximately 1/2 isosmotic, taurine efflux is not significantly increased from the isosmotic control at 60 min, and it is only 13% of that in skate RBC under the same conditions. In addition, hagfish cell volume increases immediately to 180% of control upon exposure to hypoosmotic medium, with no significant RVD over 1h period. Skate RBC under the same conditions reduce cell volume by 16% in one hour. The finding that VA taurine efflux and RVD are greatly reduced in the band 3-deficient hagfish RBC supports our hypothesis that band 3 is involved in VA taurine release.

Supported by NSF grant DCB-9102215

CELL DENSITY MEASUREMENT AS AN INDEX OF CELL VOLUME CHANGE IN RED BLOOD CELLS EXPOSED TO HYPO-OSMOTIC MEDIA.

Robert L. Preston[1], Laura P. Hartema[2] and Stephen A. Miller[2]
[1]Department of Biological Sciences,
Illinois State University, Normal, IL. 61761.
[2]Biology Department, College of the Ozarks, Point Lookout, MO. 65726

Most cells have at least some capacity to regulate their volume after exposure to hypo-osmotic media. Cells exposed to dilute medium rapidly swell, gaining water osmotically. The cells respond to this increase in volume by releasing solutes which causes the cell volume to decrease. The red cells of the marine polychaete, Glycera dibranchiata have been shown to volume regulate under these conditions by releasing amino acids (in particular, taurine) and potassium (Costa and Pierce, J. Comp. Physiol. 151:133-144, 1983). This makes physiological sense because these two solutes together comprise the majority of the intracellular solutes in these cells, each being about 200 mM.

In the present study, we have begun the initial characterization of the swelling phase and subsequent volume regulation by these cells. Earlier studies on RBCs (e.g. Costa and Pierce, J. Comp. Physiol. 151:133-144, 1983) have employed a coulter counter or capillary hemocytometry to measure cell volume changes. Both of these techniques have advantages and limitations, but in general, both require a minute or more to make detailed volume measurements. Very recently we developed a new technique employing measurement of RBC density to measure cell swelling rates in times as short as 5 sec. This will permit us to evaluate more precisely the factors (such as exposure to mercurials, etc.) which might modify the physiological behavior of cells during the initial swelling phase after exposure to hypo-osmotic media.

The objectives of this project were to verify the validity of the cell density measurement technique and apply it to evaluation of the characteristics of cell volume regulation by Glycera RBCs. An outline of the method used (which might be termed cellular sedimentation pycnometry) for a typical experiment follows: RBCs were removed from several worms, washed in isotonic seawater (NaSW, 950 mOsM) and resuspended at about 20% hematocrit. A 0.1 ml aliquot of this suspension was pipetted into a microfuge tube and centrifuged for 5 sec at 10,000 x g in a microfuge. The supernatant was removed and 0.3 ml hypo-osmotic NaSW was added. After exposure to this medium for a fixed time interval (ranging from 5 sec to 4 h), triplicate 0.1 ml aliquots were placed into microfuge tubes containing 0.5 ml hypo-osmotic medium and 0.5 ml of a mixture of phthalate acid diethyl ester (DEP) and phthalic acid bis (3,5,5-trimethylhexyl) ester (TMHP). These tubes are then centrifuged for 5 sec at 10,000 x g in a microfuge.

If the RBCs were less dense than the DEP/TMHP mixture they remained in the aqueous (NaSW) upper layer. If the cells were more dense than the DEP/TMHP mixture, they passed through the lower phthalate layer and formed a pellet. In some cases cells were present in both layers. The relative number of cells in the lower layer was measured by removing the aqueous and phthalate media and measuring hemoglobin content with 1.0 ml Drabkin's reagent. After a reaction period of at least 2 hr, hemoglobin content was evaluated by reading the absorbance of the extract at 540 nm in a spectrophotometer. The densities of the DEP/TMHP mixtures were precisely calibrated using a microbalance and glass pycnometers that were previously calibrated against water. The densities of DEP and TMHP alone were determined to be 1.1144 ± 0.0003 g/ml (23.5°C) and 0.9659 ± 0.0005 g/ml (23°C) respectively (n = 6). Thus, mixtures of these two liquids could be made to obtain any density between these two values. Control experiments have shown that DEP and TMHP are relatively inert physiologically and exposure to these phthalates for periods of up to 1 hr seems to have no effect on cell stability or transport.

Glycera RBCs were exposed to 50% NaSW (diluted 1:1 with deionized water) and 100% NaSW for 5 min and then the cell density was determined by cellular sedimentation pycnometry (Fig 1). The change of the cells from the lower phase to the upper phase was quite sharp but occasionally intermediate distributions were observed. Therefore, the data were graphically extrapolated to the density at which 50% of the cells were present in both upper and lower layers (Fig 1).

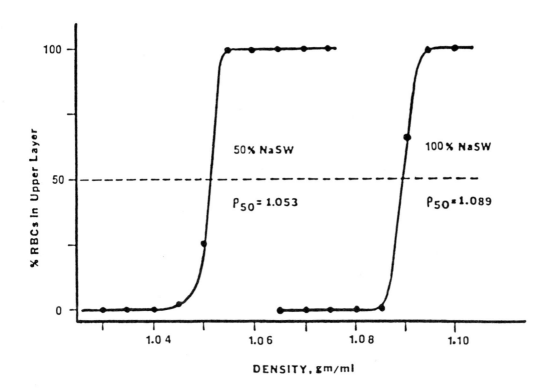

Fig 1. Glycera RBCs were incubated in 100% NaSW or 50% NaSW for 5 min at 23°C and then transfered to microfuge tubes containing phthalate medium of different densities. These tubes were then centrifuged at 10,000 x g for 5 sec. The cell density was estimated by the extent to which the RBCs penetrated through the phthalate medium to form a pellet. Relative cell number was estimated by using Drabkin's reagent to measure hemoglobin content spectrophotometrically. The apparent density at which 50% of the RBCs penetrate the lower phthalate layer (p_{50}) was 1.089 g/ml in 100% NaSW and 1.053 g/ml in 50% NaSW.

This value (p_{50}) was taken as the measure of RBC density. In 100% NaSW p_{50} = 1.089 g/ml. After 5 min in 50% NaSW p_{50} = 1.053 g/ml. Assuming that in 100% NaSW 75% of the weight of the RBCs is water (Preston, unpublished data; Machin and O'Donnell, J. Comp. Physiol. 117:303-311, 1977) and assuming that the water component of the cell doubles in reaching osmotic equilibrium with 50% NaSW, then the calculated cell density for Glycera RBCs should be 1.049 g/ml. This value is quite close to the observed value of 1.053 g/ml. Other experiments using media at concentrations intermediate between 100% and 50% NaSW show similar agreement with predicted and observed density values. This suggests that under the conditions employed for this experiment cell density is a reasonably accurate index of cell volume change. These data are also consistent with the hypothesis that these cells behave as nearly perfect osmometers during the early phases of cell swelling and regulation. Additional experiments have shown that this method can resolve cell density differences of up to at least 0.002 g/ml. At this level of precision careful temperature control is necessary since the thermal coefficients of expansion for DEP and TMHP may cause apparent differences in phthalate densities of about 0.001 g/ml per degree centigrade.

The timecourse of cell swelling was measured using 1.06 g/ml phthalate medium. In 50% NaSW the density of Glycera RBCs becomes less than 1.06 g/ml in about 20 sec. This density is about 80% of the final value (1.053 g/ml) reached at equilibrium. These data may be used to calculate an apparent water permeability coefficient. Depending on a number of assumptions regarding membrane surface area, etc. the apparent permeability to water in these cells is about 1/3 to 1/10 that of mammalian RBCs. This observation is consistent with the data of Machin and O'Donnell (J. Comp. Physiol 117:303-311, 1977) who report similar differences in water permeability. These data are also consistent with estimates of solute leak permeabilities in marine invertebrate tissues that suggest that these tissues, in general, have much lower leak permeabilities than mammalian tissues (Preston, in Kinne, RKH (ed), Comparative Aspects of Sodium Cotransport Systems. Comp. Physiol. Basel, Karger, 7:1-129, 1990). Preliminary experiments using cellular sedimentation pycnometry have shown that osmotically induced water flux is temperature sensitive (approx. Q_{10} = 1.4). It has also been shown that pretreament of the RBCs with ethanol increases the swelling rate in 50% NaSW.

In conclusion these data have demonstrated that cellular sedimentation pycnometry provides a very simple and fast method for estimating cell density changes in response to osmotic dilution. Measurements can be made at exposure times as short as 5 sec. Quantitatively, at least in the initial phases of swelling in response to hypo-osmotic media, the extent of cell density change is consistent with cell volume changes caused by osmotic water flow. This method should have considerable utility in future studies evaluating the effects of various agents on the cellular responses to osmotic stress.

(This work was supported by NIEHS grant 2P30ES0382-04 and an American Heart Association, Maine Affiliate Grant-in-Aid. Stephen A. Miller and Laura P. Hartema were selected as PEW Fellows supported by the PEW Foundation).

EFFECT OF HYPO-OSMOTIC MEDIA ON TAURINE FLUX IN POLYCHAETE RED BLOOD CELLS.

Robert L. Preston[1], Sarah J, Janssen[1], Karen S. Peterson[1],
Kristin A. Simokat[2] and Kristi L. McQuade[3]
[1]Department of Biological Sciences,
Illinois State University, Normal, IL. 61761.
[2]Southwest Harbor, Maine, 04679
[3]Department of Chemistry, McKendree College, Lebanon, Il. 62254

In a series of studies in this laboratory, we have shown that taurine transport by the hemoglobin containing coelomocytes (red blood cells, RBCs) of the marine polychaete, Glycera dibranchiata, is readily inhibited by low concentrations of mercuric chloride (Chen and Preston, Bull. Environ. Contam. Toxicol. 39:202-208, 1987; Preston and Chen, Bull. Environ. Contam. Toxicol. 42:620-627, 1989; Preston et al., Bull. MDIBL 26:129-132, 1990). The influx of taurine is inhibited 50% after a 1 min exposure to seawater containing 20 uM mercuric chloride. We have concluded that the probable site of action of mercuric chloride is the membrane transport carrier for taurine or some closely associated moiety. However, mercuric chloride is not specific for this carrier since it is highly reactive with sulfhydryl groups present in most proteins. To evaluate the other effects of mercuric chloride in these cells we have also studied mercurial inhibiton of D-glucose transport in Glycera RBCs (Preston et al., Bull. MDIBL 30:51-53, 1990) and have begun a series of studies on the effects of mercury on cell volume regulation by these cells.

In dilute media, the red cells of Glycera display a regulatory volume decrease (RVD) mediated, in part, by the release of amino acids (in particular, taurine) and potassium (Costa and Pierce, J. Comp. Physiol. 151:133-144, 1983). These two solutes are present at high concentrations in the RBCs, each being about 200 mM. The data presented here describe some of our preliminary results on the effect of osmotically dilute media on the influx and efflux of taurine by Glycera RBCs.

The basic techniques used in this study have been described in detail elsewhere (Chen and Preston, Bull. Environ. Contam. Toxicol. 39:202-208, 1987). Briefly, the transport of ^{14}C-taurine was measured using an artificial seawater medium (NaSW) at 12°C. In some experiments, the Na in the medium was replaced on an equimolar basis with choline Cl to provide a Na free seawater (CSW). The RBCs were separated from the incubation medium by centrifuging the cells through dibutylphthalate (DBP). Influx (5 min incubation period) was measured in RBCs exposed to NaSW or CSW diluted up to 50% with deionized water. The RBCs were pre-equilibrated with the dilute medium 5 min before influx measurements were made. In some experiments, influx was measured in the presence of the competitive inhibitors, beta-alanine, gamma-aminobutyric acid and hypotaurine. Efflux measurements were made by preloading the RBCs with ^{14}C-taurine for 60 min, washing the cells in 100% NaSW and then exposing the labelled RBCs to 100% or 50% NaSW for various time periods. The amount of ^{14}C-taurine in the aqueous upper layer after the RBCs were centrifuged through DBP was taken as a measure of taurine released by the cells.

Taurine influx in media ranging from 50% to 100% NaSW did not change (Table 1). This result was somewhat unexpected in that earlier experiments

(Preston and Chen, Bull. Environ. Contam. Toxicol. 42:620-627, 1989) have shown that as NaCl in NaSW was replaced with choline Cl influx dramatically decreased to about 25% of the control value in 100% NaSW. Assuming that influx would decrease solely in response to decreased Na concentration in 50% NaSW the predicted influx should be less than 50% of the value in 100% NaSW. The fact that there is no significant change in the rate of influx suggests that some compensatory stimulation of influx may be occurring activated by low osmotic pressure. Taurine influx was also measured in CSW diluted from 100% to 50% on the same batch of RBCs (Table 1). The Na-independent component of taurine influx approximately doubled as the medium is diluted from 100% to 50% CSW. Deducting the Na-independent flux from the Na-dependent flux, it is apparent that the Na-dependent component has decreased about 50% and the Na-independent component has doubled (Table 1). These data are consistent with the presence of a volume activated component to taurine flux that has been observed elsewhere (e.g. Fincham, et al. J. Membrane Biol. 96:45-56, 1987).

Table 1: Taurine Influx in Hypo-osmotic Seawater

Medium	Percent Seawater Concentration	Influx $\pm$ S.E. (n = 3) ($\mu mol \cdot min^{-1} \cdot l \cdot cell\ water^{-1}$)	J_{dil}/J_{con}	p
NaSW	100	214 $\pm$ 2	-	
	90	213 $\pm$ 10	0.99	N.S.
	70	211 $\pm$ 4	0.98	N.S.
	50	197 $\pm$ 16	0.92	N.S.
CSW	100	67 $\pm$ 3	-	
	90	75 $\pm$ 1	1.12	<0.1
	70	107 $\pm$ 1	1.60	<0.001
	50	126 $\pm$ 2	1.88	<0.001
Na Dependent	100	147		
Component	90	138		
(NaSW-CSW)	70	104		
	50	71		

Glycera RBCs were pre-equilibrated in the media above for 5 min and then incubated with 1 mM ^{14}C-taurine for 5 min. J_{dil}/J_{con} = ratio of taurine influx in the dilute medium to the influx of taurine in 100% NaSW or 100% CSW. Student's t-test was used to compare the statistical significance (p) of the experimental conditions to the control. N.S. = no statistical significance.

Competition experiments in 100% and 50% NaSW and CSW indicated that there was little significant change in the ability of the competitors beta-alanine, hypotaurine and gamma-amino butyric acid to inhibit influx. For example, the ratio (J_I/J_o) of ^{14}C-taurine influx in the presence (J_I) and absence (J_o) of beta-alanine was 0.10 (100% NaSW), 0.10 (50% NaSW), 0.08 (100% CSW) and 0.07 (50% CSW). This suggests the selectivity of the influx pathway in dilute media resembles that in control medium. These data contrast with that of Fincham et al. (J. Membrane Biol 96:45-56, 1987) in fish erythrocytes that show a decrease in selectivity for substates in dilute media.

The logical role for this Na-independent influx component in these cells might be as a pathway for taurine efflux during RVD. To directly address this possibility, we measured efflux from RBCs preloaded with ^{14}C-taurine and then exposed to 100% and 50% NaSW and CSW for 60 min (Table 2). The data indicate that taurine efflux is stimulated in 50% NaSW compared with 100% NaSW (49%) and in 50% CSW compared with 100% CSW (175%). These data are consistent with the hypothesis that the swelling of Glycera RBCs hypo-osmotic medium activates a volume sensitive Na-independent pathway that may be involved in RVD in these cells. As an operational hypothesis, we propose that this efflux pathway may be identical with or similar to the Na-independent influx pathway for taurine. Further work is being conducted to confirm or disprove this possibility.

Table 2: Taurine Efflux in Hypo-osmotic Seawater

Medium	Percent Seawater Concentration	Efflux $\pm$ S.E. (n = 3) (CPM$\cdot$h$^{-1}\cdot$cell vol^{-1})	J_{dil}/J_{con}	p
NaSW	100	1540 $\pm$ 224	-	
	50	2296 $\pm$ 59	1.49	<0.05
CSW	100	1313 $\pm$ 52		
	50	3606 $\pm$ 370	2.75	<0.005

Glycera RBCs were preloaded with 0.1 mM ^{14}C-taurine for 5 min in 100% NaSW. The RBCs were then washed in 100% NaSW and then transferred to control or dilute seawater as indicated above. Efflux is expressed as CPM. The total CPM in an aliquot of cells equivalent to that used for the efflux measurements was about 120,000 CPM. J_{dil}/J_{con} = ratio of taurine influx in the dilute medium to the efflux of taurine in 100% NaSW or 100% CSW. Student's t-test was used to compare the statistical significance (p) of the experimental conditions to the control. N.S. = no statistical significance.

(This work was supported by NIEHS grant 1P50ES0382-05 and an American Heart Association, Maine Affiliate Grant-in-Aid. Karen Peterson was the recipient of a MDIBL Scholarship and Kristin Simokat was a recipient of a Hearst Fellowship).

CELL SWELLING CAUSES INCREASED DIDS BINDING IN ERYTHROCYTES OF THE LITTLE SKATE, RAJA ERINACEA

Thad R. Leffingwell[1], Mark S. Musch[2], and Leon Goldstein[3]
[1]Southwestern College, Winfield, KS, [2]Univiversity of Chicago, Chicago, IL, [3]Brown University, Providence, RI

Erythrocytes of the little skate respond to hypotonic swelling by releasing taurine, a free amino acid maintained at high intracellular concentrations. The taurine transport system is readily inhibitable by DIDS, as well as other Band-3 inhibitors, incdicating a possible involvement of Band-3 in the taurine efflux system (Goldstien and Brill, Am. J. Physiol. 260: R1014-R1020, 1991). Previous experiments have shown an increased affinity of Band-3 for H2DIDS in hypotonic media (460 mOsm) (Flanagan, et al. Bull. MDIBL, 30:75, 1991). The purpose of the present experiments was to repeat the hypotonic binding experiments and extend the experimental conditions tested to include 660 mOsm (a milder hypotonic stress), as well as ethylene glycol EIM (elasmobranch incubation medium), and NH_4Cl EIM, which have been previously shown to cause isosmotic swelling of the erythrocytes.

Blood was drawn from the skate and the erythrocytes were immediately isolated by centrifugation. The cells were brought up to 2% hematocrit in either 940 mOsm (control isotonic), 460 mOsm, 660 mOsm, ethylene glycol, or NH_4Cl EIM. Five ml of the RBC suspension was then transferred immediately to a flask for incubation with 5 uCi of 3H2DIDS (0.5 uM, sp. act. 2 Ci/mmol) in a shaking water bath at $16^{\circ}C$. Three 150 ul aliquots were removed from the incubations at 0,5,10,15,30, and 60 min. and transferred into 1.5 ml microcentrifuge tubes containing 850 ul of the corresponding EIM with 0.5% albumin. The samples were washed twice with 1000 ul of the albumin EIM to remove all unbound or reversibly bound 3H2DIDS. The samples were then solubilized with 400 ul of a 1:1 soluene, isopropanol mixture, transferred to scintillation vials, and decolorized with 200 ul of 30% H_2O_2. The 3H2DIDS bound to the cells was then determined by liquid scintillation counting in Hionic-fluor (Packard). A separate experiment with tritiated polyethylene glycol (m.w.=4000) showed that essentially none of the radioactivity in the medium was trapped in the pellet.

In the hypotonic media, 460 and 660 mOsm, 3H2DIDS binding at 60 min. was increased nearly 2-fold and 1.5-fold respectively. In the isosmotic media, a 2-fold increase of binding at 60 min. was seen with the ethylene glycol EIM but no increased binding was observed with the NH_4Cl EIM. The increases in 3H2DIDS binding were similar at 0.1 mM and 0.5 uM 3H2DIDS. Earlier published reports had shown that the affinity of Band-3 for H2DIDS was increased in hypotonic media (Flanagan, 1991). Our results, however, show no significant difference in initial rate of binding (affinity) in any of the conditions. We attribute this difference to the use of a higher specific activity 3H2DIDS, which has been shown by gel electrophoresis to bind more specifically to Band-3. The results do suggest an increased number of binding sites available for H2DIDS in 460, 660, and ethylene glycol EIM. This can be explained by either an increased number of Band-3 proteins being made available or a conformational change in each Band-3 which results in more available binding sites on each protein. The isosmotic results match data from taurine efflux experiments in which NH_4Cl caused only a 100% increase in efflux while ethylene glycol resulted in a 300% increase (Goldstein and Brill, J. Exp. Zool., 1990). However, the match is not as good when comparing 460 vs. 660 mOsm. Binding is increased only 50% more by 460 mOsm but taurine efflux is increased about 10 times more (Goldstein and Brill, 1991). This difference must be attributable to another factor which may regulate the use of the available transporters, and not just to the number of transporters available. These results support the hypothesis that Band-3 is involved in the taurine efflux response of skate erythrocytes to hypotonic and isosmotic swelling.

Supported by Pew Fellowship to T.R. Leffingwell and NSF grant to L. Goldstein.

VOLUME REGULATORY MECHANISMS IN HEPATOCYTES FROM RAJA ERINACEA: IMPAIRMENT BY MERCURY

N. Ballatori[1], A.T. Truong[1], J. Gardner[2] and J.L. Boyer[2]

[1]Department of Biophysics, Environmental Health Sciences Center, University of Rochester School of Medicine, Rochester, NY 14642

[2]Department of Medicine and Liver Center, Yale University School of Medicine, New Haven, CT 06510

Skate hepatocytes exposed to hypotonic media swell in proportion to the decrease in extracellular osmolarity, but subsequently exhibit a regulatory volume decrease (RVD). In contrast to hepatocytes from other vertebrate species, RVD in skate hepatocytes is associated with the release of only a small fraction of intracellular K^+, but a substantial fraction of intracellular taurine, an amino acid found in relatively high concentrations (65 mM) in skate hepatocytes (Ballatori and Boyer, Am. J. Physiol., in press). RVD in skate hepatocytes is essentially abolished by pretreatment with 50 µM mercuric chloride (Ballatori et al. Toxicol. Appl. Pharmacol. 95:279, 1988). To further characterize the mechanisms involved in RVD and in mercury's ability to impair volume recovery, the present study used primary suspension cultures of skate hepatocytes (Smith et al., J. Exp. Zool. 241:291, 1987). Intracellular water space was determined as the difference between the ^3HOH and ^{14}C-inulin distribution spaces, and ^{14}C-taurine fluxes measured by a rapid centrifugation procedure (Ballatori and Boyer, Am. J. Physiol. 254:R801, 1988).

The effects of metabolic poisons and sulfhydryl reagents on RVD is shown in Table 1. As previously observed with mercury, 2,4-dinitrophenol, iodoacetate plus KCN, N-ethylmaleimide and diamide all inhibited RVD in skate hepatocytes. In contrast, 2 mM ouabain had no effect on volume recovery (Table 1). This concentration of ouabain has previously been shown to effectively inhibit $^{86}Rb^+$ uptake into skate hepatocytes (Ballatori et al., Biochim. Biophys. Acta 946:261, 1988). Thus, RVD is not directly influenced by Na-K-ATPase activity, but appears to be dependent on metabolic energy and reduced sulfhydryl groups.

Table 1. Effects of metabolic inhibitors, sulfhydryl reagents and ouabain on RVD.

| | Relative Volume (%) | | | | |
| | Minutes after dilution with 40% H_2O | | | | |
	1	5	20	40	60
Control	157±4	148±4	136±2	134±2	129±3
2,4-Dinitrophenol	156±6	154±6	149±6	151±7*	153±6*
Iodoacetate + KCN	158±9	160±8	155±8*	159±8*	160±8*
N-Ethylmaleimide	158±8	159±9	162±8*	164±7*	167±7*
Diamide	154±7	154±6	150±7	149±12	153±11*
Control	154±4	146±8	128±6	123±5	120±4
Ouabain	153±5	147±5	132±7	128±4	124±2

Cells were preincubated for 30 min with 0.5 mM 2,4-dinitrophenol (n=6), 1 mM iodoacetate plus 1 mM KCN (n=4), 1 mM N-ethylmaleimide (n=4) or untreated (n=6), before dilution with 40% H_2O. Osmolarity was decreased from ~940 to 564 by addition of H_2O. Ouabain (2 mM, n=7) was added 2h before hypotonic challenge. Values are means ±SE. *Significantly different from control, $p<0.05$, with Student t test.

Because both K$^+$ and taurine contribute to RVD, additional studies examined the effects of mercury on intracellular K$^+$ content and taurine fluxes. Skate hepatocytes treated with 50 μM mercury lost 42% of intracellular K$^+$ over 2h (from 0.416 to 0.242 μEq/mg protein), whereas intracellular Na$^+$ increased from 0.156 to 0.571 μEq/mg protein over the same time interval. Because the gain in Na$^+$ is more than double the K$^+$ lost, Na$^+$ entry probably contributes to both the mercury-induced cell swelling and its inhibition of RVD. In addition, mercury pretreatment inhibited the rapid initial release of taurine normally observed after cell swelling. The inability to release this osmolyte may also contribute to the observed inhibition of RVD.

The reversibility of mercury's effects on RVD was examined by adding the sulfhydryl-containing compounds glutathione (GSH) and dithiothreitol (DTT) at various times after mercury addition (Table 2). DTT was able to nearly completely reverse the effects on RVD even when added as late as 3 min after mercury administration. In contrast, GSH blocked the effects of mercury only when added simultaneously with the toxic metal. GSH addition at 1 or 3 min post mercury only partially reversed the inhibition of RVD (Table 2). Because DTT enters cells relatively easily, whereas exogenous GSH is largely excluded from the intracellular space, these observations indicate that mercury is interacting with intracellular components to elicit its effects on RVD.

Table 2. Dithiothreitol (DTT) and glutathione (GSH) reverse the mercurial inhibition of RVD.

| | Relative Volume (%) | | | |
| | Minutes after dilution with 40% H$_2$O | | | |
	1	10	30	60
Control	156±1	141±2	126±4	118±4
DTT, 0.5 mM	159±3	144±2	128±3	128±4
HgCl$_2$, 0.05 mM	159±2	165±7	175±13	188±20
HgCl$_2$ plus DTT at t=0 min	158±5	142±4	129±7	120±4
HgCl$_2$ plus DTT at t=1 min	162±3	146±2	129±3	120±2
HgCl$_2$ plus DTT at t=3 min	163±5	150±5	138±2	124±3
Control	157±1	143±3	125±2	118±2
GSH, 0.5 mM	156±3	140±3	124±3	115±1
HgCl$_2$, 0.05 mM	159±1	163±5	173±10	186±15
HgCl$_2$ plus GSH at t=0 min	155±3	143±3	129±4	117±6
HgCl$_2$ plus GSH at t=1 min	163±2	162±2	159±5	157±5
HgCl$_2$ plus GSH at t=3 min	161±2	162±5	159±4	158±9

Cells were diluted with 40% H$_2$O at time zero. Where indicated, this H$_2$O also contained HgCl$_2$ to give a final concentration in the cell suspension of 50 μM. DTT (n=3) and GSH (n=4) were added at either 0, 1, or 3 min. Values are means ±SE.

These findings suggest that RVD in skate hepatocytes is an energy and sulfhydryl-dependent process. Mercury appears to be interacting with intracellular components, and presumably sulfhydryl groups to elicit its inhibitory effects on RVD.

Supported by National Institutes of Health Grants DK39165 (NB), DK34989 (JLB) and ES03828 to the Mount Desert Island Biological Laboratory.

ION TRANSPORT IN RED CELLS OF ATLANTIC MACKEREL (SCOMBER SCOMBRUS): EFFECT OF CATECHOLAMINES AND VOLUME PERTURBATIONS

Thomas J. McManus and Lawrence C. Starke
Division of Physiology, Duke Univ. Medical Center, Durham, NC 27710

Cell volume and pH regulatory systems in the red cells of many fish have been shown to play a significant role in adaptational strategies used by these animals to cope with environmental changes. Most of the work in this field has been on red cells from freshwater adapted rainbow trout (Oncorhynchus mykiss). In these cells, binding of catecholamines to plasma membrane β_1-adrenoceptors activates an amiloride–sensitive Na/H exchange that helps maintain an alkaline internal pH in the face of stress–induced extracellular acidosis. An alkaline cytosol promotes the high hemoglobin–oxygen affinity necessary for efficient uptake at the gill, particularly under hypoxic conditions. A high internal pH also prevents an acid–induced unloading of oxygen from hemoglobin (the Root effect). One consequence of this response, however, is red cell swelling secondary to net uptake of Na via Na/H exchange .

In the saltwater environment, marine teleosts face serious osmoregulatory challenges. The ever-present tendency for salt loading is circumvented by a delicately balanced relationship between seawater imbibition (Smith, Quart. Rev. Biol. 7:1-26, 1932) and salt extrusion at the gill (Evans, in Fish Physiology, Ed. by Hoar and Randall, Vol. XB, Chapter 8, 1984). Any disturbance in this relationship along with continued drinking can produce a marked increase in plasma Na and Cl. We have described just such an event in stressed Atlantic mackerel (Scomber scombrus) (Starke and McManus, Bull. MDIBL, 28:17–19, 1989). Following capture (hook and line) and confinement to live cars, there was a marked rise in plasma osmolality due primarily to an increase in Na and Cl. Presumably, salt extrusion at the gill was inhibited by stress-related factors. Surprisingly, in reponse to this hypertonic challenge, the red cells did not shrink, but maintained constant volume by taking up a precise osmotic equivalent of both ions. After a week of confinement, both extracellular osmolality and intracellular Na returned to normal. In view of the work on rainbow trout cited above, it appears likely that the mechanism of net salt uptake by the red cells during the stress response was catecholamine activation of Na/H exchange. The purpose of the experiments reported here was to test this assumption in vitro, and to explore further the volume regulatory characteristics of mackerel red cells.

Mackerel caught in Frenchman Bay were confined to live cars immersed in seawater. After the stress response was established (24-48 hours), blood was drawn from the caudal vein using a heparinized syringe. A gauze sponge soaked in seawater and wrapped round the gills prevented suffocation. The red cells and plasma were separated by centrifugation and washed 3 times in an iced NaCl solution isosmotic with the hypertonic plasma (470 mOsm/kg). Resuspended in a similar hypertonic Ringer to a final hematocrit of about 15%, the cells were incubated overnight at 4°C to achieve steady states with respect to ions and water. The Ringer contained (mM): 2.5 KCl, 0.75 CaCl_2, 5 glucose, 1 MgCl_2, 10 TMA-TES (TMA = tetramethylammonium; TES = (N-Tris [hydroxymethyl] methyl-2-aminoethane-sulfonic acid, titrated to pH 7.7 at 12°C with TMA hydroxide), and sufficient NaCl to bring the final osmolality to 470 mOsm/kg. After washing and resuspending the cells in this same Ringer, all subsequent incubations were performed at 12°C.

Osmolality was determined using a Wescor vapor-pressure osmometer (5100B). Na and K were determined on perchloric acid (3.6% PCA/1.5 mM CsNO_3) extracts of packed cells using an IL flame photometer. Cl was measured with a Radiometer CMT-10 chloridometer. Ion and water content of the cells was calculated from these data in combination with wet weight/dry weight cell water determinations (Schmidt and McManus, J. Gen. Physiol. 70: 59-79, 1977). To study Na/H exchange directly, net proton efflux from DIDS-treated cells into unbuffered media was assayed by monitoring external pH changes as previously described (Payne and McManus, Bull. MDIBL, 28: 57–59, 1989).

Addition of norepinephrine (10^{-5} M) to the incubation medium promoted cell swelling during the first hour associated with a marked accumulation of Na. After the first hour, the increase ceased and volume remained stable for six hours. Ouabain had no effect on either the rate or magnitude of swelling, although it did change the relative proportions of Na and K in the cells. The sum of Na and K contents at any given point in time, however, was not altered by the presence of ouabain.

To confirm that this effect was due to catecholamine activation of Na/H exchange, net proton efflux was monitored directly by the method referred to above. Before hormone was applied, the unbuffered medium acidified slowly at a rate of –0.05 pH units/minute. For the first 30–40 seconds after addition of 1 μM isoproterenol, there was little change in rate of acidification, but then an abrupt increase to –0.35 pH units/minute occurred. It is interesting to note that we observed a similar delay in onset of proton efflux after activation of Na/H exchange by the phorbol ester, PMA, in shark red cells (Payne and McManus, op. cit.), as did Grinstein in PMA–stimulated lymphocytes (PNAS, 82: 1429–1433, 1985). In contrast to these results, as well as those reported by Motais in experiments on trout red cells (Progress in Cell Research. 1: 179-193, 1990), we were unable to show any effect of phorbol esters in this system. This could be related to the intracellular pH, since Motais et al. (op. cit.) showed that PMA stimulation of Na influx occurred only when internal pH was below 7.6, whereas isoproterenol was effective up to pH 8.0. Therefore, a complete study as a function of internal pH will be necessary before a definite conclusion can be reached concerning the effect of PMA on these cells. Amiloride (0.5 mM) promptly blocked acidification of the external medium, further supporting the conclusion that the catecholamine activated Na/H exchange.

Cell shrinkage also stimulated external acidification. After recording a similar control rate of –0.05 pH units/minute, medium osmolality was abruptly increased to 650 mOsm/kg by addition of a small volume of concentrated NaCl. The rate doubled to –0.1 pH units/minute, but again only after a 30–40 second time delay. Amiloride also blocked this process. When the osmotic increase was produced by adding concentrated KCl, there was no change in the control rate of acidification, but subsequent addition of concentrated NaCl (with no change in final osmolality) produced an increased rate of acidification <u>without</u> a time delay. Thus, the net efflux of protons required external Na, but the time delay was related only to the hypertonic stimulus itself. These experiments confirm that both cell shrinkage and catecholamine addition stimulate Na/H exchange in mackerel red cells.

Swelling induced ion transport was examined by observing net K loss from cells incubated in a hypotonic medium. Since intracellular Na was already elevated due to the hypertonic stress response, net loss of cell water could also occur by the well known dehydrating capability of the Na pump (Clark et al. Biochim. Biophys. Acta 646: 422-432, 1981). To avoid this complication, cells were depleted of Na by overnight incubation in a Na free, high K medium (470 mOsm/kg), then washed and reincubated in the presence of ouabain in a Na–free hypotonic medium (200 mOsm/kg) in which Na was replaced by TMA. One batch of cells was washed with a solution in which Cl was replaced by sulfamate, a permeant anion that has been shown to have little effect on the volume and pH of red cells (Payne, Lytle and McManus. Am J. Physiol. 259: C819–C827, 1990). They were then re-incubated in a similar Cl–free hypotonic medium. The swollen cells showed a net loss of K sensitive to bumetanide (10^{-3} M) and inhibited by removal of Cl, suggesting the presence of swelling induced [K–Cl] cotransport.

When norepinephrine (10^{-5} M) was added to cells suspended in the hypotonic medium, net K efflux was inhibited about 40%, which cannot be related its effect on Na/H exchange since there was no Na present, either in the cells or the medium. We have shown that catecholamine, or cyclic–AMP, addition directly inhibits [K–Cl] cotransport in the duck red cell (Starke and McManus, J. Gen. Physiol. 92: 42a-43a, 1988). If that is also the case in the mackerel, then there may be a coordinated regulation between the swelling and shrinkage induced pathways — agents which stimulate one coordinately inhibit the other — similar to the effect we have previously demonstrated in both duck (Starke and McManus, op. cit.) and dog (Parker et al. J. Gen. Physiol. 96:1141-1152, 1990) cells. Further experiments will be required to confirm this conclusion.

The fish red cell is a useful model in this regard since it manifests [K–Cl] cotransport like many other red cells, has a shrinkage–induced Na/H exchange like the dog cell, but is hormonally regulated like [Na–K–2Cl] cotransport in the duck cell. Thus, it offers the investigator a unique opportunity to explore the interaction between cell volume dependent and hormonally controlled transport systems that share no ions in common, and therefore probably represent distinct gene products that are regulated by a common mechanism.

The initial phase of this work was supported in part by a pilot study grant from the Center for Membrane Toxicity Studies, MDIBL. Dr. Starke's present address is Department of Physiology and Biophysics, Univ. of Texas Medical Branch, Galveston, TX 77550

THE RESPONSE OF THE ACTIN CYTOSKELETON OF DOGFISH (SQUALUS ACANTHIAS) RECTAL GLAND CELLS TO HYPOTONICITY, HIGH POTASSIUM AND MERCURIALS

Richard M. Hays[1], Yuri Natochin[2],
Rimma Parnova[2] and Arnost Kleinzeller[3].
[1]Department of Medicine, Albert Einstein College of Medicine,
Bronx, New York 10461
[2]Sechenov Institute of Evolutionary Physiology and Biochemistry,
St. Petersburg, USSR
[3]Department of Physiology, University of Pennsylvania,
Philadelphia PA 19104

Hypotonic stress of dogfish shark (Squalus acanthias) rectal gland cells produces rapid cell swelling followed by a regulatory volume decrease. In addition, a disappearance of F-actin at the basolateral membrane of rectal gland cells was observed by fluorescence microscopy of rhodamine phalloidin labeled F-actin over the 5 - 10 minute period of cell swelling (Ziyadeh et al., Amer. J. Physiol., in press). This was followed by a gradual recovery of the F-actin organization after a 30 - 60 minute period, during the regulatory volume decrease. High medium K^+ reduced basolateral fluorescence (A.Kleinzeller and J.W. Mills, Biochim.Biophys.Acta 1014:40,1981), and organic mercurials reduced overall F-actin fluorescence in the rectal gland cell. (A.Kleinzeller et al., Biochim. Biophys. Acta 1025,21,1990).

We have asked whether the changes in basolateral F-actin are accompanied by a reduction in F-actin in the entire cell. We employed the rhodamine phalloidin binding assay to measure whole cell F-actin in slices of rectal gland which were incubated either in standard isotonic elasmobranch incubation medium (900 mosM/kg H_2O) or in hypotonic medium in which 160mM NaCl was removed, giving a final osmolality of approximately 600mosM/kg H_2O. High K^+ incubation medium, in which Na was equivalently replaced by K^+ was used in a second series of experiments. In addition, the effects of p-chloromercuri benzene sulfonate (pCMBS) and phenylmercuric acetate (pMA) were determined. For the rhodamine phalloidin binding assay (Ding et al., Amer.J.Physiol. 260,C9, 1991), slices were weighed, then incubated at $15^{\circ}C$ for varying time periods in control or hypotonic medium, fixed with 3.7% formaldehyde in calcium-free dogfish medium containing 0.1mM EGTA, and then incubated for 1 hour in calcium-free dogfish medium containing 0.3uM rhodamine phalloidin. The slices were then washed, the fluorescence extracted with methanol, and read in a spectrofluorimeter. Results were expressed as fluorescence per mgm tissue wet weight, and in all cases, the ratio of the test sample to the simultaneously determined control sample was determined, (Table 1).

Table 1:Effect of Hypoosmolality, High K^+ and Mercurials on F-Actin

Time (min)	Hypotonicity	High K^+ % of Control	pCMBS	pMA
10	100±14 (3)[*]			
60	99± 8 (2)			
120		97±12 (3)	60±7 (3)[**]	27 (1)

[*]   number of experiments   [**]   p< 0.02

At 10 minutes exposure to hypotonic medium, there was no change in the F-actin content compared to the control. This was also the case at 60 minutes incubation. In 4 rectal glands, the affect of high K^+ was determined at 120 minutes. Again, there was no effect on total cell F-actin, although, as previously reported (Kleinzeller and Mills, loc. cit.) the basolateral pool of F-actin is considerably diminished by high K^+. In contrast $10^{-3}M$ pCMBS significantly decreased F-actin at 120 minutes. $10^{-4}M$ pCMBS decreased F-actin by 15% at 120 minutes in a single experiment. $10^{-3}M$ pMA decreased F-actin by 73% in a single experiment.

Our results show that overall cellular F-actin is not decreased by hypotonicity or high K^+, despite the decrease in basolateral F-actin as seen by fluorescence microscopy. This suggests that changes in actin cytoskeleton following hypotonic stress or high K^+ are regional in nature, with basolateral depolymerization presumably being offset by apical polymerization or polymerization elsewhere in the cell. It is also possible that actin filaments are significantly shortened during hypotonicity and high K^+ and yet retain the ability to bind rhodamine phalloidin, but are not readily seen in the conventional fluorescence microscope.

Supported by NIH grant DK AM 03858, the Soros Foundation (Dr. Parnova) and The MDIBL NIEHS Center for Membrane Toxicity Studies (Drs. Natochin and Parnova).

OSMOLARITY AND CELL VOLUME CHANGES OF CHLORIDE CELLS: THE NATURE OF THE RAPID SIGNAL FOR ADAPTATION TO SALINITIES OF FUNDULUS HETEROCLITUS

Jose A. Zadunaisky[1], Susan Cardona[1], Dawn Roberts[2],
David Giordano[3], E.J. Cragoe, Jr.[4], and Kenneth Spring[5]
[1]Department of Physiology and Biophysics, New York University
Medical Center, New York, NY 10021,
[2]Ogelthorpe University, Atlanta, GA 30319,
[3]Bridgewater College, Bridgewater, MA 02325,
[4]Consultant, Nacogdoches, TX 75963,
[5]Kidney and Elecrolyte Laboratory, NIH, Bethesda, MD 20892

The adaptation from low to high salinities of euryhaline fish as in the case of Fundulus heteroclitus occurs in two distinct stages. There is a quick adaptation that consists of rapid increase of chloride secretion by the gill in general (Maetz, J., Bornancin, M., 1975. Fortschr. Zool. 23:322). The second adaptation is a long term one, taking several days or weeks and is manifested by an increase in the electrical resistance and formation of new filaments in the tight junctions, an increase in the number of chloride cells and the presence of more sites for ouabain binding in the chloride cells, an index of greater number of molecules of the NaK ATPase (Sardet, C. et al. 1979. J. Cell Biol., 80:96, Karnaky, K.J., et al. 1976, J. Cell Biol. 70:157). The information concerning the mechanisms for the slow, secondary adaptation is more clear than the explanation for the rapid changes that must occur in euryhaline species to survive changes in osmolarity and salinity that can represent, at times, a quick transition from few milliosmoles to the 1000 milliosmoles of full seawater.

It is known that during adaptation to higher salinities euryhaline fish drink sea water, absorb the salts through the intestine and secrete them through the gill, utilizing this organ in preference to the kidney (Smith, H.W. 1930 Am. J. Physiol. 93:485). The changes in osmolarity and salinity then occur not only in the outside or apical side of the chloride secretory cells but also in the basolateral side of the epithelium. In fact, there is a transient increase in the concentration of NaCl in the plasma of euryhaline fish that can reach up to 50 mM above the normal level and last for 12 to 20 hours (Holmes and Donaldson: L. Fish Physiol. 1:30: 1969). As the gill secretory epithelium starts to secrete NaCl, the plasma concentration is reduced and the second phase, under the control of hormones, possibly thyroxine and cortisol (Burdick, C.J. 1985 Biol. Bulletin. 169:559; Foskett, J.K., et al. 1983 J. Exp. Biol. 106:225) induces the changes in protein synthesis and cell proliferation. In some species, such as Fundulus heteroclitus of the North Atlantic the chloride cells remain in the gills and its adjacent areas, namely the epithelium of the roof of the mouth and the opercular epithelium, in fresh water adapted specimens.

In other species such as the eel, tilapia, or the trout, the chloride cells are atrophied in fresh water adapted fish and only a few remaining cells are found (Zadunaisky, 1984 Fish Physiol., Vol.XB).

We have studied the importance of osmolarity and its effect on the volume and secretory properties of chloride cells of Fundulus heteroclitus. The purpose was to establish if the rapid cell volume changes that can occur during the transition from low to high osmolarity coincide with changes in the transport rate or permeability of the chloride cells.

The preparation utilized was the opercular epithelium of <u>Fundulus</u> <u>heteroclitus</u> if the North Atlantic, adapted to fresh or sea water. The methodology consisted of the use of Ussing chamber experiments which indicate the chloride secretion by measurements of the short circuit current and in the use of quanititative optical microscopy. Optical sections and reconstruction methods based on the use of refractive optics (Nomarski) were utilized to ascertain the cell volume of the chloride cells in different conditions.

<u>Fundulus</u> <u>heteroclitus</u> trapped in fluvial estuaries draining into Frenchmans Bay in Maine were immobilized by pithing, the opercular epithelium was dissected out and mounted in highly modified Ussing chambers (MDIBL Bul. Vol. 30, 1991, pp. 58-59). The short circuit current (SCC) was measured and transepithelial electrical resistance was determined by sending small pulses of voltage and reading the displacement in the SCC. For the optical experiments, opercular epithelia were mounted in a perfusion chamber, placed on the stage of an optical microscope, and observed through diffraction optics, with the use of a television camera and monitor plus a compact disc image storage system. Appropriate computer programs permitted the optical sectioning of chloride cells at intervals of 1.5 microns, subsequent tracing and calculation of the surface area of each section, and finally the determination of the cell volume (Furlong, T.J., Spring, K.R., 1990. Am. J. Physiol. 258,C1016-C-1024). Osmolarity changes were induced with mannitol.

In the cell volume experiments, we found that a 20% increase in the osmolarity of the apical side produced the predicted and expected change in cell volume decrease of chloride cells. Subsequent return to normal osmolarity of the Ringer's solution produced the expected cell volume increase, with return to the initial cell volume in fresh water adapted Fundulus. However, the response occurred only in fresh water adapted operculi. In tissues obtained from sea water adapted specimens we found no change in volume with the same exposure to 20% increased in osmolarity with mannitol.

The volume response from the apical side in fresh water adapted specimens and the lack of response in the sea water adapted ones is an indication of a drastic change in water permeability of the apical membrane of the chloride cells from one condition to the other.

Fresh water adapted opercular membranes produced none or very small electrical potential difference (0 to 0.1 mV) and an osmotic load in the apical side promoted a very slow, gradual increase in the potential difference. (2 mV in 1.5 hrs)

Shrinkage of chloride cells by the addition of 50 milliosmoles of mannitol to the basolateral side produced a rapid, sustained increase of the SCC in seawater adapted fish. The dose response curve of the normalized increase in SCC against osmolarity in 6 complete curves showed a saturable event with a maximum of 220% increase at 300 mOsm and 1/2 maximal stimulation at 100 mOsm. (Fig. 1)

The activation of the chloride secretion at 50 mOsm on the basolateral side of seawater adapted fish was tested after the addition of specific agents that affect the $2Cl^-Na^+K^+$ cotransporter, the Na^+/H^+ exchanger, the Cl^-/HCO_3^-

3nd the Ca^+/Na^+ exchangers.

The use of Bumetanide at $10^{-5}M$ produced a slow reduction in SCC as expected from its inhibition of the $2Cl^-Na^+K^+$ cotransporter. Delivery of 50 mOsm to the basolateral side failed to elicit any response indicative of activated chloride secretion. (Fig. 2)

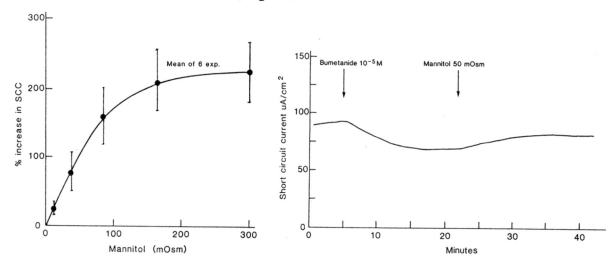

Fig. 1: Dose response curve for increases in osmolarity with mannitol on the basolateral side of opercular epithelia of _Fundulus h._ Note the saturable curve obtained for six experiments. Bars indicate standard error of the mean.

Fig. 2: Lack of stimulation of the chloride current of an isolated operculum of _Fundulus h._ after treatment with the inhibitor bumetanide. A slow reduction in current is stopped by the increase in 50 mOsms, however, there is no increase in the current.

In the previous summer, (MDIBL Bul. Vol. 30 1991 pp. 58-59), it was demonstrated that a Cl^-/HCO_3^- exchanger exists in the basolateral membrane of Fundulus operculi. Inhibition of the Cl^-/HCO_3^- exchanger with $10^{-4}M$ DIDS (4,4'-Diisothiocyanatostilbene-2,2'-disulfonic Acid) drastically reduces the SCC, as does the absence of HCO_3^-. HCO_3^- appears to be essential for maintenance of the Cl^- secretion, as pH is an important component of the signaling mechanism for Cl^- transport. However, in 8 experiments inhibition of the Cl^-/HCO_3^- exchanger with $10^{-4}M$ DIDS had no appreciable effect on the hypertonicity response when 50 mOsmols of mannitol was administered to the basolateral side. The recovery from the effects of the DIDS alone was 75% of the initial SCC (Fig. 3)

The Na^+H^+ antiport is known to have a significant role in both pH and cell volume regulation. High concentrations of both Amiloride and its more specific derivatives, 5-(N,N-hexamethylene) Amiloride and 5-(N-ethyl-N-Isopropyl) Amiloride, at $10^{-2}M$ respectively, were shown in 14 experiments to significantly inhibit or obliterate the hypertonicity response to mannitol altogether. Activation of the Na/H antiport can be considered as one of the primary responses to a change in osmolarity. (Fig 4)

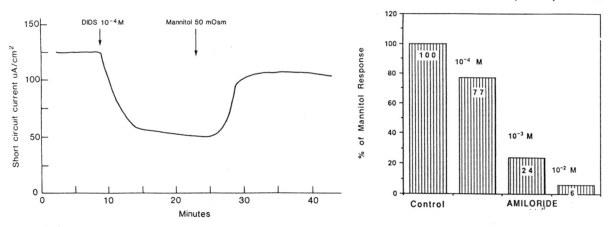

Fig. 3: Complete response to an increase in basolateral osmolarity after blocking the Cl^-/HCO_3^- exchanger with DIDS. Apparently there is no involvement of this exchanger in the increase in chloride secretion produced by hypertonicity on the basolateral side of opercular epithelia of <u>Fundulus h.</u>

Fig. 4: The inhibition of the mannitol response by different doses of Amiloride, in isolated opercular epithelia of <u>Fundulus h.</u> The maximal dose of 10^{-2} produces complete inhibition. Analogs of Amiloride produced inhibition at lower concentrations, (see text).

Based on previous unpublished research, Ca^{++} has emerged as part of the rapid signaling mechanism in this system. Preliminary results (Fig. 5) from several experiments which explore the effects of Ca^{++} free ringer and 100 uM Ca^{++} ringer on the hypertonicity response indicate a relationship between intracellular Ca^{++} and Cl^- secretion. At "0" Ca^{++} the SCC response to the osmotic challenge is extremely low. At a concentration of 100 uM Ca^{++}, the response is better, however, still lower than the control.

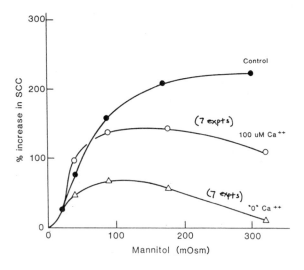

Fig. 5: Dose response curve for increases in osmolarity with mannitol on opercular epithelia of <u>Fundulus h.</u> in Ca^{++} free ringer and 100 uM Ca^{++} ringer.

To summarize, the events following the increase in plasma osmolarity during the adaptation to seawater and the subsequent cell shrinkage involve activation of both the $2Cl^-Na^+K^+$ cotransporter and the Na/H antiport. The increased intracellular Cl^- and the effect of pH may be responsible for the increased chloride secretion through the apical membrane. The need for intracellular Ca^{++} mobilization, perhaps for Cl^- channel activity, is also essential for the cell shrinkage response.

Acknowledgements: This work was supported by NIH grant EY01340 to JAZ. Dawn Roberts and David Giordano are grateful to the Grass Foundation (DR) and the Pew Foundation (DG) for fellowships that permitted them to participate in this research.

A PUTATIVE MUSCARINIC RECEPTOR (M4) FROM THE CEREBELLUM OF THE SPINY DOGFISH, <u>SQUALUS ACANTHIAS.</u>

J.P. Schofield[1], J. Rock[2], J.Appel[3], and J. N. Forrest Jr.[4]
[1]MRC Molecular Genetics Unit, Hills Rd, Cambridge, England.
[2]College of the Atlantic, Bar Harbor, Maine 04609.
[3]Duke University, Durham NC 27710.
[4]Yale University School of Medicine,333 Cedar St, New Haven CT 06510.

Previous studies have demonstrated that muscarinic receptor subtypes are expressed in specific regions of the central nervous sysytem. The M1 subtype has been localised to the cerebral cortex and corpus striatum, whilst the M2 subtype is expressed in the medulla-pons. The M4 receptor has a similar distribution to the M1 subtype, and one proposal is that they are particularly involved in the cognitive processes of learning and memory (Nathanson, N.M., <u>Ann. Rev. Neurosci.</u> 10:195-236) . As cognitive processes rely upon synaptic plasticity the receptor subtype heterogeneity may be a mechanism by which such plasticity can be achieved (Braun, <u>et al</u>, <u>Biochem. Biophys. Res. Comm.</u> 149:125-132, 1987). A single cell could vary its expression of several receptor subtypes, and thus control its response to a neurotransmitter. We have sought to apply the polymerase chain reaction (PCR) to isolate a dogfish cerebellar cDNA encoding a putative seven transmembrane domain G-coupled muscarinic receptor. The new sequence shows strong homology to the rat muscarinic M4 receptor, which has previously been shown by <u>in situ</u> hybridization to be abundant in the granule cell layer of the cerebellar cortex (Braun, <u>et al</u>, 1987). No muscarinic receptors could be identified by amplification of rectal gland cDNA, confirming previous physiological investigations.

RNA was extracted from the dogfish shark brain and rectal gland (1g of each organ). Tissues were immediately flash-frozen in liquid nitrogen upon removal from the animal to prevent ribonuclease action. Samples were polytron homogenized in lysis buffer and mRNA isolated (Fast-track mRNA extraction kit,InVitrogen,San Diego CA). First strand cDNA was synthesized from 1µg of mRNA (RT-PCR kit, InVitrogen). Approximately one tenth of the first strand cDNA was amplified with degenerate muscarinic receptor primers A and B (figure 1). The amplification profile was 35 cycles each of: dissociation at 95°C for 0.5 min, annealing at 30°C for 0.5 min, and extension at 60°C for 0.5 min. Control amplifications omitting input cDNA were simultaneously performed to exclude reaction contamination. Following the primary PCR, a second-round PCR was conducted on 1µl of the 50µl primary product using internal ("nested") muscarinic primers C and D (figure 1). The cycling profile was as for the first round PCR but limited to 20 cycles. No products were obvious following the primary amplification, yet a single 1kb band was clearly visible in brain cDNA after the secondary PCR (figure 2). Identical experiments performed using dogfish rectal gland cDNA consistently failed to amplify any products.

<u>Figure 1</u>: Conserved muscarinic receptor amino acid sequences within transmembrane domains (TM) II and VII to which redundant primers were designed.

```
        Sense primer A:              5'.TVNNYY.3'
                                         TM II

        Sense primer C:              5'.CADLII.3'
                                         TM II

        Anti-sense primer B:         3'.CYALCN.5'
                                         TM VII

        Anti-sense primer D:         3'.GYWLCY.5'
                                         TM VII
```

<u>Figure 2:</u> Agarose gel of the second round amplification of dogfish cerebellum (C), and rectal gland cDNA (R)..A single band of ~1Kb is amplified only from cerebellar cDNA.

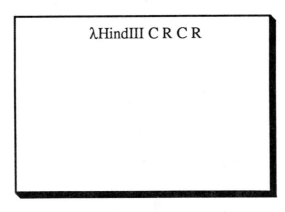

λHindIII C R C R

The single amplified product was cloned into a plasmid vector (TA-cloning kit, InVitrogen), and transformed into competent <u>E. coli</u> cells. Recombinants were screened by blue/white selection, and mini-preparation DNA prepared for DNA sequencing. A modified alkaline lysis double-stranded sequencing protocol was used along with Sequenase 2.0™ enzyme and chain terminators (USB, Cleveland OH). Sequence information was analysed on an Apple Macintosh II using DNA STAR (LASERGENE™). By aligning the translated sequence with known muscarinic receptors from other species, the receptor appears most similar to the M4 type found in rats and humans. Using these methods, only partial sequence data is generated. To obtain the complete sequence of this new muscarinic receptor, future work will include screening a dogfish genomic library.

<u>Figure 3:</u> Amino acid alignment between the translated shark cerebellar putative muscarinic receptor and the rat M4 muscarinic receptor. Asterisks (*) indicate amino acid complete homology. Transmembrane domains (TM) are underlined.

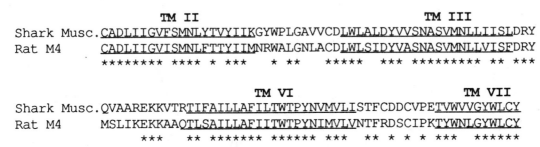

```
                    TM II                            TM III
Shark Musc.CADLIIGVFSMNLYTVYIIKGYWPLGAVVCDLWLALDYVVSNASVMNLLIISLDRY
Rat M4     CADLIIGVISMNLFTTYIIMNRWALGNLACDLWLSIDYVASNASVMNLLVISFDRY
           ******** **** * ***   * **  *****  *** ********* ** ***


                          TM VI                      TM VII
Shark Musc.QVAAREKKVTRTIFAILLAFILTWTPYNVMVLISTFCDDCVPETVWVVGYWLCY
Rat M4     MSLIKEKKAAQTLSAILLAFIITWTPYNIMVLVNTFRDSCIPKTYWNLGYWLCY
           ***    ** ******* ***** ***  ** * * * *** ******
```

Within a short period of time (5-6 days) we were able to identify a new muscarinic receptor in the shark cerebellum and determined much of its sequence. We were also able to confirm what physiological evidence has suggested, that there are no muscarinic receptors expressed in the dogfish shark rectal gland. The reverse genetics techniques used proved to be highly efficient both in mimimizing the time required to obtain results and in conserving the number of animals used. We required only a single shark to obtain 1 gram samples from the brain and rectal gland.

This work was supported by fellowships from the Lucille P. Markey Trust and Blum-Halsey Fellowships (JPS), a Fellowship from the Pew Foundation (JR), and the American Heart Association (Maine Affiliate) (JA,JNF). We aknowledge the kind gifts of molecular biology reagents from United States Biochemicals, Cleveland OH, and InVitrogen Corporation, San Diego CA.

A PAIR OF NOVEL ORPHAN RECEPTORS ARE EXPRESSED IN THE RECTAL GLAND OF THE SPINY DOGFISH, SQUALUS ACANTHIAS.

J.Paul Schofield[1], D. Stephen Jones[1], and John N. Forrest[2], Jr.
[1]MRC Molecular Genetics Unit, Hills Road, Cambridge CB2 2QH, England
[2]Yale University School of Medicine, 333 Cedar St., New Haven CT 06510

The rectal salt gland of the spiny dogfish is a hollow tubular epithelial gland situated on the posterior abdominal wall. Analogous to the mammalian kidney ascending loop of Henlé, it has become established as a powerful physiological model system for the study of salt transport mechanisms. Adenosine analogues have been shown to be potent stimulators of ion transport when perfused into isolated glands. The target receptor(s) however have not been identified in the dogfish. Recent molecular biological techniques using the powerful polymerase chain reaction (PCR) and redundant oligonucleotide primers have identified cDNAs encoding both stimulatory and inhibitory adenosine receptors from the dog thyroid gland. The sequences of these cDNAs when translated into amino acid residues are members of the protein superfamily of G-coupled receptors. Prior to their expression and performance of binding studies the natural ligands for these receptors are unknown, and are therefore termed "orphan". Characteristically they possess a similar tertiary structure of seven hydrophobic transmembrane (7TM) spanning regions. We predicted that the adenosine receptors within the rectal gland will have a similar structure, and will provide an ideal system for subsequent detailed molecular physiological studies. In addition, this approach has the potential to identify all expressed cDNAs encoding G-coupled receptors within the rectal gland. For example, the human vasoactive inhibitory peptide (VIP) receptor has been cloned and shown to be a 7TM G-coupled receptor. VIP perfused into the rectal gland has been shown to be active in stimulating water and ion secretion, and is probably active via an homologous receptor.

Immediately following collection a single rectal gland was flash frozen by dropping into liquid nitrogen. This was in order to inhibit the activity of endogenous ribonucleases. Poly-A[+] messenger RNA (mRNA) was prepared (Fast-Track™, InVitrogen San Diego CA), and first strand cDNA reverse transcribed to serve as template for amplification (RT-PCR kit, InVitrogen San Diego CA). A programmable thermocycler (Techne PHC-2™, Princeton NJ) was used to amplify regions of G-coupled receptors using a pair of degenerate oligonucleotide primers designed to hybridise to segments of the third and sixth transmembrane domains. An amplification profile for 35 cycles was: dissociation at 95°C for 0.5 min, annealing at 40°C for 0.5 min, and annealing at 60°C for 1 min. A slow ramp rate was programmed between annealing and extension to maximise stability of the redundant primer extension products. Taq polymerase enzyme was from Cetus, Norwalk CT. The reaction products ranging in size from 200bp to 1Kb were separated in a 1% low melting agarose gel, and purified onto glassmilk (USBioclean™, USB Cleveland OH). Individual products were TA-cloned (Marchuk et al, Nucl. Acids Res. 19, 1154,1991; TA-cloning kit, InVitrogen, San Diego CA) and recombinants screened utilising blue-white selection (Ullmann et al, J. Mol. Biol. 24, 339-343, 1967). Double-stranded plasmid DNA sequencing use Sequenase 2.0™ enzyme (USB, Cleveland OH).

Sequence manipulation utilised the DNAStar computer program (Lasergene™) running on an Apple Macintosh IIcx microcomputer. A single open-reading frame continued from the translated 3'end of the IIITM primer for two clones of about 450bp. Alignment with each other showed a 50% amino acid absolute identity. Homology was greatest at the predicted IV and V transmembrane regions. However, database comparison with Genbank revealed only weak homology to the rat substance K receptor sequence, again at predicted IVTM and VTM predicted segments. There is no significant homology with the published sequences of dog adenosine stimulatory (RCD8) or inhibitory (RCD7) receptors (Libert et al, Science 244, 569-572, 1989; figure 1).

Figure1: Amino acid alignment of shark rectal gland orphan receptors (SOR1, SOR2), rat substance K receptor (RATSUBK), dog adenosine receptors (RCD8, RCD7), rat testis adenosine receptor (TGPCR1), rat muscarinic M1 receptor (RATM1), rat serotonin HT1 receptor (RAT5HT1), and human vasoactive inhibitory peptide receptor (HUMVIP). Asterisks (*) denote highly conserved residues between receptors. The amino acids of the fourth transmembrane domain (TM IV) are underlined.

```
SOR1      VVVAYPIRQRIRPRSCAYIVAF--IWLVSIGVS-MPSSLHT
              *                    *           *
SOR2      VVVAYPIRQRITLSCCGLIMG--AIWVLSMALAPQPPSTSC
              *                    *           *
RATSUBK   MAIVHPFQPRLSAPSTKA-IIAGI-WLVALALAS-PQC-F-YS
              *                    *           *
RCD8      IAIRIPLRYNGLVTGTRAKGIIAVCWVLSFAIGLTPM--LGWN
              *                    *           *
RCD7      LRVKIPLRYKTVVTPRRAAVAIAGCWILSFVVGLTPL--FGWN
              *                    *           *
TGPCR1    LRVKLTVRYRTVTTQRRIWLFLGLCWLVSFLVGLTPM--FGWN
                                   *           *
RATM1     FSVTRPLSYRAKRTPRRAALMIGLAWLVSFVLWA-PAILF-WQ
              *                    *           *
RAT5HT1   VAIRNPIEHSRFNSRTKAIMKIAIVWAISIGVSV-PIPVIGLR
              *                    *           *
HUMVIP    LSITYFTNTPSSRKKMVRRVVCILVWLLAFCVSL-PDTYYLKT
                                 TM IV
```

The pair of putative orphan 7TM receptor encoding cDNAs isolated from the dogfish rectal gland were used as probes to screen a rectal gland cDNA library. No positives were detected either on screening this library or another available rectal gland library provided by Dr. E. Benz. This may reflect a very low abundance message, undetectable by conventional cDNA library screening. We are proceeding to construct a shark total genomic DNA library, and plan to screen this with the orphan receptors.

It remains an open question as to whether the sequences of the two orphan receptors are those coding for dogfish stimulatory and inhibitory receptors. Even though the sequences are markedly different from the adenosine receptor of "higher" species, we have previously shown that the dogfish atrial prepronatriuretic factor (prepro-CNP) exhibits poor homology with the equivalent cDNA sequences of all other species. An alternative is that these new orphan receptors may be the shark VIP or substance K receptors for example. These questions will be resolved following isolation and expression of the full-length cDNA sequences in conjunction with ligand-binding studies, i.e. exactly the route taken to identify canine RCD7 and RCD8 as adenosine receptors.

This research was sponsored by fellowships from the Lucille P. Markey Charitable Trust to J.P.S. and D.S.C.J., a Blum-Halsey Fellowship to J.P.S., and from the American Heart Association (Maine Affiliate) to J.N.F. We gratefully aknowledge the kind gifts of molecular biology reagents from United States Biochemicals, Cleveland OH, and InVitrogen Corporation, San Diego CA.

A MOLECULAR ANALYSIS OF MUSCARINIC RECEPTOR SUB-TYPES EXPRESSED WITHIN THE AORTA OF THE SPINY DOGFISH, SQUALUS ACANTHIAS.

J.P. Schofield[1], I. Lombardo[2], and D.H.Evans[3].
[1]MRC Molecular Genetics Unit, Hills Road, Cambridge, England.
[2]Yale University School of Medicine, 333 Cedar St, New Haven CT 06510.
[3]University of Florida, Department of Zoology, Gainesville FL 32611.

Toxicity studies on isolated aortic rings from the spiny dogfish identified a differential contractile response to cadmium, selectively inhibited (50%) by atropine, a potent inhibitor of muscarinic receptors (Evans & Weingarten, Toxicology 61:275-281, 1990; Evans et al., Toxicology 62:89-94, 1990). There had been no previous documentation of cholinergic innervation of Squalus vasculature. As the tissue specificity of receptor sub-types may reflect an underlying functional variation (e.g. adenyl cyclase inhibition or stimulation of phosphatidylinositol turnover) a greater understanding of the site of action of cadmium would be achieved by further molecular dissection of this contractile response. The aim of this work was both to establish the presence and to perform a molecular characterisation of the sub-types of muscarinic receptors in the aorta. Highly sensitive and specific methods had previously been developed to amplify muscarinic receptors using the polymerase chain reaction (PCR, and see article by JPS in this Bulletin and Bulletin 1990).

Aortic tissue (about 1g) was quickly dissected from several dogfish, and immediately flash-frozen in liquid nitrogen to prevent endogenous ribonuclease action. The tissue was polytron homogenized in lysis buffer and mRNA isolated (Fast-track mRNA extraction kit, InVitrogen, Can Diego, CA). First strand cDNA template for amplification was synthesized from about 1μg of aortic mRNA (RT-PCR kit, InVitrogen). Two rounds of muscarinic receptor amplifications were performed using degenerate primers, according to established protocols (see accompanying article by JPS in this Bulletin for details). No products were obviously amplified following the first round of PCR, as had been previously observed with Squalus cerebellum and rectal gland mRNA. However, a second round amplification using nested internal primers resulted in three clear products of the predicted length (around 800-1200bp). These were visualised by loading one-tenth of the reaction volume onto a 0.8% agarose gel containing ethidium bromide. A control reaction was simultaneously performed, excluding input cDNA, and was negative through both rounds of amplification. Future work will involve cloning these amplified products and DNA sequencing to establish them as muscarinic receptor encoding cDNAs as well as their sub-types. Full-length sequence for expression studies may be obtained by using these sequences as molecular probes to hybridize against a Squalus acanthias genomic DNA library.

In this short report we have described our initial experiments aimed at providing a more detailed molecular understanding of the target sites of action of the heavy metal cadmium. Our results suggest that muscarinic receptors are present in Squalus aortic tissue, confirming the physiological prediction. This knowledge will also facilitate the future development of more general molecular approaches to the investigation of the specific mechanisms of action of other toxic compounds in marine organisms.

JPS was supported by Fellowships from the Lucille P. Markey Charitable Trust, and a Blum Halsey Fellowship. IL was supported by an American Heart Association medical student research fellowship (Connecticut Affiliate). DHE was supported by NSF Grant DCB 8916413 and NIH EHS-P30-ESO3828 to the Centre for Membrane Toxicity Studies.We aknowledge the kind gifts of molecular biology reagents from United States Biochemicals, Cleveland OH, and InVitrogen Corporation, San Diego, CA.

CONSTRUCTION AND SCREENING OF DOGFISH (SQUALUS ACANTHIAS), HAGFISH (MYXINE GLUTINOSA), AND TOADFISH (OPSANUS TAU), HEART AND BRAIN cDNA LIBRARIES.

D. Stephen C. Jones[1], J. Paul Schofield[2] and J.N Forrest Jr[3]
[1] Dept of Obstetrics & Gynaecology, Cambridge University. Cambridge. England
[2] MRC Molecular Genetics Unit, Hills RD. Cambridge. England.
[3] Yale University Medical School, 333 Cedar St. New Haven CT 05610

Understanding in almost all fields of biological research is currently being advanced rapidly by the application of recombinant DNA technology to the processes under investigation. One of the key methods in this advance is that of cDNA cloning and sequencing. This permits the identification and characterization of genes (and therefore proteins) without the necessity of purifying the protein. For example the recent identification of the Squalus cardiac CNP by the authors required only a single animal. The aim of this project was to extend this work by the construction of cDNA libraries from the hearts and brains of the hagfish Myxine glutinosa, and the toadfish Opsanus tau as well as the rectal gland from Squalus acanthias. These were then to be screened in order to isolate cDNAs encoding natriuretic factors, the homologues of oxytocin and vassopressin and novel 7-transmembrane domain receptors.

Hagfish and toadfish hearts (8 and 4 respectively) and brains (10 and 6) were dissected from anaesthetised animals and immediately frozen in liquid nitrogen. A single rectal gland was similarly frozen after removal from a dog fish. Messenger RNA was prepared using a Fast Track kit from InVitrogen. Double stranded cDNA was synthesized and after adapter ligation, was size fractionated by preparative agarose gel eletrophoresis. This entire process yielded approximately 100ng of purified cDNA. This was then ligated to the λgt10 vector (which had previously digested with EcoRI). After in vitro packaging the resulting bacteriophage were plated on the selective E.coli strain, C600 Hfl. The background in each case was less than 3% and the total number of recombinants shown in the table below:

Hagfish brain	1.3×10^5
Hagfish heart	2.6×10^5
Toadfish brain	5.0×10^5
Toadfish heart	1.3×10^5
Dogfish rectal gland	5.5×10^5

These libraries were then screened with several probes: The previously isolated squalus CNP, and two oligonucleotides encoding the highly conserved amino acid motif KLDRIG. The latter resulted in 2 positively hybridizing regions from the toadfish heart library which are currently being characterised in Cambridge. The rectal gland library (and also another similar library provided by Dr Benz) was screened with the two orphan receptor probes (see the Bulletin article by JPS). These screens produced no positively hybridizing clones indicating that the mRNA corresponding to these receptors is rare in rectal gland tissue.

DSCJ and JPS received fellowships from the Lucille P. Markey Charitable Trust and JPS also from the Blum-Halsey Trust. JNF is in part supported by the American Heart Association (Maine Affiliate). We also acknowledge the kind gifts of reagents from United States Biochemicals, Cleveland OH and Invitrogen Corporation San Diego CA.

MECHANISMS OF CADMIUM TOXICITY IN <u>RAJA</u> <u>ERINACEA</u> ELECTRIC ORGAN

Oliver M. Brown and John S. Andrake
Departments of Pharmacology and Pediatrics
State University of New York Health Science Center
Syracuse, New York 13210

We have been studying the mechanisms of evoked neurotransmitter release and the mechanisms of action of cadmium (Cd) on excitable membranes. As a model system in this effort, we are using the electric organ of the skate. This preparation allows us to quantitate and correlate changes in physiology, neurotransmitter release, and ion movement in the same preparation.

The skate, <u>Raja erinacea</u>, has weak electric organs in the tail, one on each side of the spinal cord; the organ probably serves the fish as a system of communication. This organ has been characterized as being a purely cholinergic tissue, utilizing the neurotransmitter, acetylcholine (ACh). Our previous studies indicate that the electrophysiological and biochemical properties of <u>Raja</u> electric organ provide for a unique and useful system with which to examine synaptic events and the mechanisms of cadmium toxicity. We continue to study the effects of Cd and other agents on <u>in vitro</u> sections of this unique and homogeneous model nervous system (for background, please see our reports in the previous four MDIBL Bulletins, and FASEB J. $\underline{3}$:A890, 1989). The evoked electrical discharge of the tissue was monitored as a means of assessing its physiological status. Following loading with ^{3}H-choline (^{3}H-Ch), synaptic release of ^{3}H-ACh was evoked by electrical stimulation or by high concentrations of potassium (K). Ch and ACh in the tissue and released from the tissue were measured by liquid scintillation counting of tritium-labeled samples and by thin layer chromatography. Calcium uptake by the tissue was stimulated by electrical pulses, by high concentrations of K, or by application of ACh, and was measured by liquid scintillation counting of ^{45}Ca.

<u>Raja</u> tissue, stimulated and incubated in buffer containing tritiated choline, readily took up ^{3}H-Ch and converted it to ^{3}H-ACh, which was released upon subsequent electrical stimulation. We recently found that ^{3}H-ACh release was also stimulated by brief incubation of loaded tissue in buffer containing 70 mM K (high K) (Fig. 1). Also, we found that our ability to measure ACh release was markedly enhanced by treating the tissue with the cholinesterase inhibitor, neostigmine (100 μM). Further, we verified by thin layer chromatography the identity of the stimulation-released tritium as ^{3}H-ACh. Cadmium inhibited the release of ^{3}H-ACh evoked by both electrical pulses and high K in a concentration-related fashion. This effect of Cd, as well as its inhibition of electrical discharge, was overcome in part by increasing extracellular Ca concentration. These findings are consistent with those of others, and with the hypothesis that Cd blocks cell membrane Ca channels or other Ca mechanisms.

To further test the interactions between Cd and Ca we examined the effects of Cd and other agents on voltage-dependant uptake of ^{45}Ca in the isolated electric organ (see Andrake, et al., MDIBL 30:91-93, 1991). The robust ^{45}Ca uptake seen in electrically or high K stimulated <u>Raja</u> tissue was eliminated in the presence of 100 μM Cd. We also found that d-tubocurarine, an antagonist of nicotinic ACh receptors, blocked Ca uptake (and evoked discharge), but did not affect ^{3}H-ACh release, suggesting that most of the ^{45}Ca uptake measured is post-synaptic. The L-type Ca channel antagonist, nifedipine, blocked ^{45}Ca uptake and evoked potential without affecting ^{3}H-ACh release, indicating that the post-synaptic, but not the pre-synaptic Ca channels have L-type properties. In contrast, nickel, which is characterized as a T-type Ca channel antagonist, did block ACh release. This property is shared by cadmium and verapamil, suggesting that the effects of nickel, cadmium, and verapamil on electric organ are (at least in part) to block pre-synaptic T-type Ca channels. The block of ACh release by cadmium, nickel, and verapamil would be expected to result in and an inhibition of post-synaptic ^{45}Ca uptake by preventing stimulus-response coupling;

although an additional direct effect of these agents on post-synaptic Ca channels cannot be excluded by these data. However, our preliminary experiments with bath-applied ACh (10 μM) demonstrated a vigorous uptake of ^{45}Ca with this stimulus (Fig. 2). This ACh-stimulated ^{45}Ca uptake was blocked by d-tubocurarine, but not by 100 μM Cd, which is evidence that Cd does not act post-synaptically, via L-type Ca channels, in this Raja electric organ.

We appreciate the technical help of Niamh O'Leary-Liu and Stephen Hamm. The support of the NIEHS MDIBL-CMTS (ES03828-05) and the Hearst Foundation is gratefully acknowledged.

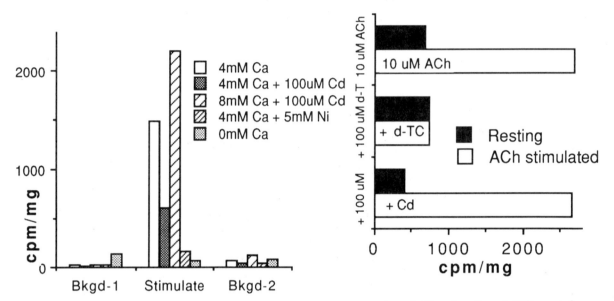

Figure 1 - High K release of ^{3}H-ACh, expressed as cpm ^{3}H per mg tissue. Bkgd-1 and Bkgd-2 are washes before and after high K incubation. In 4 mM Ca buffer, ^{3}H-ACh release was inhibited by 100 μM Cd and 5 mM Ni. Cd inhibition was antagonized by increasing Ca to 8 mM. High-K = incubation in 70 mM K buffer.

Figure 2 - ACh-stimulated ^{45}Ca uptake, expressed as cpm ^{45}Ca per mg tissue. Ca uptake was blocked by tubocurarine (d-TC, 100 μM) but not by 100 μM Cd. ACh Stim = incubation in 10 μM ACh.

THE EFFECT OF PRESYNAPTIC BLOCKADE ON CADMIUM TOXICITY TO VASCULAR SMOOTH MUSCLE OF SQUALUS ACANTHIAS

David H. Evans and Tes Toop, Department of Zoology, University of Florida
Gainesville, FL 32611

Our previous studies have determined that constriction of the ventral aorta of the dogfish shark (Squalus acanthias) can be produced by both cadmium and nickel (Evans and Weingarten, Toxicology 61: 275-281, 1990). Approximately 50% of the Cd^{2+}-induced vasoconstriction can be inhibited by atropine (Evans and Weingarten, Toxicology 62: 89-94, 1990). We hypothesized that the Cd^{2+} effect was via direct interaction with muscarinic receptors on the VSM because there has not been any description of an obvious cholinergic innervation of the shark ventral aorta (e.g., Nilsson, Autonomic Nerve Function in the Vertebrates, Springer-Verlag, 1983) and addition of atropine alone did not change aortic VSM resting tension, suggesting a lack of tonic release of acetylcholine from cholinergic neurons (Evans and Weingarten, Op. Cit., 62: 1990). Despite these negative data, it is important to determine if the Cd^{2+} induced constriction could have been mediated via interaction with presynaptic neurons because it has been shown that Cd^{2+} can induce contraction of ileal longitudinal muscle of guinea pig by stimulating the release of acetylcholine (Asai et al., Br. J. Pharmacol. 75: 561-567, 1982). To discriminate between direct interactions between Cd^{2+} and VSM muscarinic receptors and indirect actions via stimulation of the release of acetylcholine, we utilized the specific, presynaptic release inhibitor ß-bungarotoxin (ß-BuTX: e.g., Rowan et al., Br. J. Pharmacol. 100: 301-304, 1990).

Rings of vascular smooth muscle (VSM; approximately 3 mm diameter) are cut from the ventral aorta and mounted (at approximately 500 mg tension) in 10 ml of elasmobranch Ringer's solution bubbled with 1% CO_2 in air (pH = 7.8) in glass tissue-chambers as described previously (Evans and Weingarten, Op. Cit.,61: 1990). In paired experiments, utilizing two rings cut from the same piece of ventral aorta, we applied 100 nM ß-BuTX (50 X maximal effective dose in other systems (e.g., Rugolo et al., Op. Cit., 1986)) to one ring and vehicle (10 µl of 0.08 N acetic acid) to the second. $CdCl_2$ was added to each bath to generate a concentration-response curve over the range of 100 nM (11 ppb) to 100 µM (11 ppm).

The addition of BuTX did not affect the tonic tension of the isolated VSM ring, and did not inhibit the constrictory effects of the addition of Cd^{2+}. In control rings, addition of 10 and 100 µM Cd^{2+} increased the tension by $18 \pm 7.3\%$ (Mean $\pm$ S.E., N=4) and $52 \pm 18\%$, respectively; in the presence of ß-BuTX, the increases were $29 \pm 10\%$ and $53 \pm 18\%$, respectively. Our previous experiments had demonstrated that 10 µM Cd^{2+} produces an approximately 14% increase in tension and 100 µM Cd^{2+} produces a 25% increase in tension in shark aortic VSM (Evans and Weingarten, Op. Cit., 61; 1990). These data are consistent with the hypothesis that presynaptic release of acetylcholine from putative cholinergic fibers is not involved in the Cd^{2+}-induced increase in tension, but one cannot rule out the hypothesis that ß-BuTX is not an effective presynaptic toxin in this system. Only further study of the efficacy of ß-BuTX on better-described, cholinergic systems in this species will differentiate between these two hypotheses; however, to date, there are no physiological data supporting a cholinergic innervation of the branchial vasculature of elasmobranchs (Metcalfe and Butler, J. Exp. Biol. 113: 253-268, 1984; Donald, Comp. Biochem. Physiol. 90C: 165-171, 1988).

These studies were supported in part by NSF DCB 8916413 (DHE) and NIH EHS-P30-ESO3828 to the Center for Membrane Toxicity Studies.

FREEZING TOLERANCE IN <u>MYTILUS EDULIS</u>

Donald A. McCrimmon[1,2] and Jennifer Rock[3]
[1]MDI Biological Laboratory, Salsbury Cove, ME 04672
[2]Present Address: Department of Biological Sciences,
Oakland University, Rochester, MI 48309
[3]College of the Atlantic, Bar Harbor, ME 04609

On Mount Desert Island, tidal range averages 3.2 m. Sessile blue mussels, <u>Mytilus edulis,</u> living in the upper regions of the intertidal zone must endure longer periods of exposure (6-8 h) to air than their counterparts in lower intertidal zones. A challenge which all <u>Mytilus</u> face is exposure to winter ambient air temperatures which can drop as low as -30 °C for prolonged periods. Under such circumstances, exposed <u>Mytilus</u> repeatedly freeze, yet survive. Mechanisms postulated to account for the tolerance of marine invertebrates to these harsh conditions include differential movement of water, increased blood cation concentrations (Murphy, D. and Pierce, S., J. Exp. Zool. 193:313-22, 1975) and increases in the concentrations of certain amino acids, including strombine and taurine (Loomis, S., <u>et al.</u> Biochemica et Biophysica Acta 943:113-118, 1988). A shift from aerobic respiration to anaerobic metabolism is associated with these physicochemical changes (Murphy, D., J. Exp. Biol. 69:1-12, 1977a; J. Exp. Biol. 69:13-21, 1977b).

We examined the relationships between exposure to different degrees of cold, intertidal location, and duration of forced anaerobic metabolism to freezing tolerance in <u>Mytilus</u>. We also gathered data on the extent of seasonal acclimation to sub-freezing temperatures in this species. All specimens were collected from shoreline of the Mount Desert Island Biological Laboratory in Salsbury Cove, Maine.

To establish a basis for tolerance to cold-exposure, in April, 1990, we collected <u>Mytilus</u> from high intertidal (HIT) or sub-tidal (SUB) locations and exposed experimental groups (n = 30 each) for 12 h in a Freas 815 low temperature incubator to test temperatures ranging from -17 to -1 °C. The mussels next were placed for at least one hour in Living Stream units (LSU's) recirculating 7 °C natural sea water. Specimens were then removed from the LSU and allowed to sit undisturbed in air for up to 3 minutes, sufficient time for natural valve closure. The posterior adductor muscle is especially susceptible to freezing injury; its failure to contract leaves the valves gaping. Mussels whose valves sealed spontaneously or which closed and remained closed after being forced shut by hand

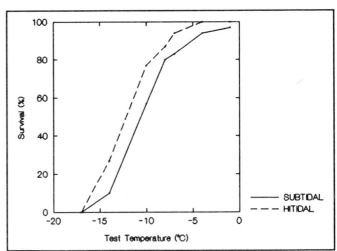

Figure 1. Survival of high intertidal and sub tidal <u>Mytilus</u> samples exposed to 12 h of sub-freezing temperatures in a low temperature incubator.

were classed as survivors. Animals whose shells did not close naturally after removal from the LSU or upon several forced closures were considered dead. The results, expressed as the percentage of each group surviving, are presented in Figure 1. HIT <u>Mytilus</u> always exhibited greater survivorship, compared to SUB (Wilcoxon signed ranks test, p < 0.03). The greatest difference in percent survival between HIT and SUB experimental groups occurred at test temperatures of -10 °C, the temperature at which a log-linear regression also estimated 50% survival between animals from the two locations.

A second series of experiments investigated in detail the process of acclimation. HIT and SUB Mytilus were placed in a LSU in which the temperature was held at a constant 13 °C for two weeks. This temperature approximated within 2 °C of the ambient seawater temperature at the time the experimental animals were first collected. Following this initial period, samples (n= 20 each) of both HIT and SUB Mytilus were removed and their freezing tolerance tested by exposure to -10 °C for 12 hours. The water temperature for the remaining mussels in the LSU was then lowered to 10 °C, where it remained for two additional weeks. Second HIT and SUB samples (n = 20 each group) were then removed and tested. The remaining mussels in the LSU were permitted to acclimate to 5 °C for a final two week period and subsequently tested. During the entire experimental period, fresh pre-cooled sea water was introduced frequently to help insure that the mussels had access to planktonic food.

For comparison with the laboratory experiments, samples of HIT and SUB Mytilus were taken also in November and December, 1990 directly from the natural environment and tested as ambient sea temperatures declined during early winter. The range of temperature variation for experimental animals was 8 °C, within 0.5 °C of that for animals acclimating at the shore. Results, expressed as percent surviving standard test conditions of -10 °C temperature exposure for 12 h, are presented in Table 1.

Table 1. Effects of 12 h exposure to -10 °C on survival of Mytilus acclimated to varying natural and experimental temperatures.

		ENVIRONMENT			LSU		
Acclimation Temp (C)		11	7.5	3.5	13	10	5
% survivorship	HIT	85	60	95	90	80	55
	SUB	40	35	85	50	45	45

Kruskal-Wallis ANOVA revealed no statistically consistent directional differences in acclimation among environmental HIT and SUB samples (x^2 = 1.77, p > 0.18). In comparison, as LSU acclimation temperatures declined, there was a significant reduction in survivorship among HIT Mytilus, while SUB animals consistently demonstrated poorer freezing tolerance, showing virtually no acclimation (x^2 = 3.97, p < 0.05). Of considerable interest to us, LSU HIT animals demonstrated apparently paradoxical acclimation. Although these HIT animals had higher overall survival, extended time in the LSU was associated with a loss of acclimation ability to decreasing temperatures. The periodic replenishment in the LSU's of fresh sea water should have eliminated nutritional stress. Therefore, we hypothesized that a lack of periodic reversion to anaerobic metabolism, which intertidal-dwelling Mytilus normally undergo during low tide, was responsible for the observed decline in freezing tolerance.

To evaluate this hypothesis, we compared survival of Mytilus held for different lengths of time under conditions requiring either aerobic or anaerobic metabolism. To avoid the potentially extreme acclimation histories of either SUB or HIT organisms, animals from mid-tidal locations were used in these experiments. Mussels were collected in April, 1991 from 3 °C sea water and divided in two experimental groups. Animals to be tested following periods of aerobic respiration were placed in a LSU circulating 3 °C aerated sea water. For comparison, animals to be tested following anaerobic metabolism were clamped shut with rubber bands, wrapped in several layers of parafilm and placed in a low temperature incubator at 3 °C. The use of parafilm and rubber bands prevented gaping and, thus, aerobic respiration during exposure to air, a possibility suggested for Mytilus by Helm and Trueman (Comp. Biochem. Physiol. 21:171-77, 1967). Samples (n = 20 each) were tested for freezing tolerance after 10, 20 and 30 h of either aerobiosis or anaerobiosis.

The significantly different (Likelihood Ratio x^2 = 19.01, p < 0.001) distributions of freezing tolerance are presented in Figure 2 and corroborate Theede et al.'s (Mar. Biol. 15:160-91, 1972) finding that Mytilus' exposure to anaerobic conditions increases freezing resistance. Our results further demonstrate that although aerobically respiring Mytilus show greater freeze tolerance after 10 h in the LSU than anaerobically metabolizing animals held for the same period in an incubator, additional time under aerobic conditions results in a marked loss of tolerance. In contrast, up to 20 h of continued anaerobic exposure increases freezing tolerance. Our results suggest, therefore, that prolonged exposure to severe metabolic stress may be adaptive to _Mytilus_ living in extreme conditions. On the other hand, loss of a stressor is quickly associated with dramatic reduction in adaptation, a phenomenon of potential significance for mussels living in sub-tidal zones, where they are only occasionally required to metabolize anaerobically. The results also emphasize the importance of providing for stimulus flux during investigations of physiological function for animals normally exposed to changing environmental variables.

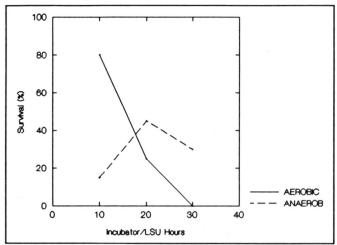

Figure 2. Tolerance of midtidal _Mytilus_ samples to varying hours of aerobic or anaerobic conditions at 3 °C and 12 h exposure to -10 °C.

This work was partially supported by NIH EHS-P30-ESO3828 to DAM and a grant from the Pew Charitable Trusts to JR.

170.

James L. Boyer, M.D.
Professor of Medicine
Chief, Div. of Digestive Diseases
Yale University School of Medicine

Gary W. Conrad, Ph.D.
Professor, Division of Biology
Kansas State University

John N. Forrest, Jr., M.D.
Professor of Medicine
Yale University School of Medicine

Freddy Homburger, M.D.
President and Director
Bio-Research Institute, Inc.
Cambridge, Massachusetts

Gregg Kormanik, Ph.D.
Associate Professor
Department of Biology
University of N.C. at Asheville

Martin Morad, Ph.D.
Professor of Physiology and Medicine
University of Pennsylvania

Robert L. Preston, Ph.D.
Associate Professor of Physiology
Department of Biological Sciences
Illinois State University

Peter F. Reilly
Executive Vice President
First National Bank of Bar Harbor

Leonard Silk
Hulls Cove, Maine
      and
Economics Columnist, The NY Times
New York, New York

S C I E N T I F I C   P E R S O N N E L   1 9 9 1

Principal Investigator	Associates
Nazzareno Ballatori, Ph.D. Assistant Professor Department of Biophysics Environmental Health Science Center University of Rochester School of Medicine	A. Truong
Edward J. Benz, Jr., M.D. Professor and Chief Hematology Section Department of Internal Medicine & Human Genetics Yale University School of Medicine	J. Appel D. Thomas K. Simokat
James L. Boyer, M.D. Professor of Medicine Director, Liver Study Unit Chief, Division of Digestive Diseases Yale University School of Medicine	M. Blumrich, M.D., Ph.D. J. Gardner D. Kinne
Oliver M. Brown, Ph.D. Associate Professor, Pharmacology Department State University of New York Health Science Center, Syracuse	J. Andrake, M.D. N. O'Leary-Liu
Gloria V. Callard, Ph.D. Professor, Department of Biology Boston University	M. Betka, Ph.D. D. Gelinas J. Haverly J. Jorgensen F. Piferrer, Ph.D.
Ian P. Callard, Ph.D. Professor, Department of Biology Boston University	L. Perez D. Schultz
Alan N. Charney, M.D. Professor of Medicine NYU School of Medicine; Chief, Nephrology Section V.A. Medical Center, NY	A. Taglietta
James B. Claiborne, Ph.D. Associate Professor, Department of Biology Georgia Southern College	E. Perry
Gary Conrad, Ph.D. Professor, Division of Biology Kansas State University	A. Conrad, Ph.D. A. Stevens

Principal Investigator	Associates
David C. Dawson, Ph.D. Professor of Physiology University of Michigan Medical School	G. Feero M. Post
James A. Dykens, Ph.D. Assistant Professor Department of Biology Grinnell College	D. Bloom D. Hoang K. May
Franklin H. Epstein, M.D. William Applebaum Professor of Medicine, Harvard Medical School; Director, Nephrology Division, Beth Israel Hospital	H. Brignull S. Hornung K. Spokes D. Wolff
David H. Evans, Ph.D. Professor, Department of Zoology University of Florida	J. Donald, Ph.D. T. Toop A. Vomachka, Ph.D.
Bliss Forbush III, Ph.D. Associate Professor of Cellular and Molecular Physiology Yale University School of Medicine	C. Forbush J. Forbush C. Lytle, Ph.D. J. Payne, Ph.D. N. Ringstad
John N. Forrest, Jr., M.D. Professor, Department of Medicine Yale University School of Medicine	C. Aller S. Aller G. Kelley, M.D. C. Kelmenson I. Lombardo D. Opdyke J. Rock
Raymond A. Frizzell, Ph.D. Prof., Dept. of Physiology and Biophysics Univ. of Alabama at Birmingham	R. Worrell
David Goldstein, Ph.D. Asst. Prof., Dept. of Biological Sciences Wright State University	S. Heflich
Leon Goldstein, Ph.D. Professor and Chairman Department of Physiology and Biophysics Brown University	J. Alley S. Brill T. Leffingwell M. Musch, Ph.D.
Richard M. Hays, M.D. Professor of Medicine Albert Einstein College of Medicine	J. Mills U. Natochin, M.D., Ph.D. R. Parnova, Ph.D.

Principal Investigator	Associates
Steven C. Hebert, M.D. Assoc. Prof. of Medicine and Physiology Brigham & Women's Hospital (Renal)	G. Gamba M. Hediger M. Lombardi A. Miyanoshita
John H. Henson, Ph.D. Assistant Professor Department of Biology Dickinson College	D. Nesbitt
D. Stephen C. Jones, Ph.D. MRC Research Fellow MRC Molecular Genetics Unit Cambridge, England	
Karl Karnaky, Jr., Ph.D. Associate Professor Department of Anatomy and Cell Biology Medical University of South Carolina	M. Kennedy A. McCraw D. Nelson J. Stidham, Ph.D. K. Suggs
George W. Kidder III, Ph.D. Professor and Chairman Department of Biological Sciences Illinois State University	B. Haltiwanger
Rolf Kinne, M.D., Ph.D. Director, Max-Planck-Institut fuer Systemphysiologie Dortmund, West Germany	M. Hulsewah R. Moore B. Schoelermann
Evamaria Kinne-Saffran, M.D. Senior Investigator, Max-Planck-Institut fuer Systemphysiologie Dortmund, West Germany	A. Nies
Arnost Kleinzeller, M.D., Ph.D., D.Sc. Professor Emeritus, Department of Physiology University of Pennsylvania	G. Booz, Ph.D. F. Ziyadeh, Ph.D.
Thomas J. Koob, Ph.D. Research Assistant Professor, Department of Biology University of New Mexico	C. Monian J. W. Straus, Ph.D. J. Trotter, Ph.D.
Gregg A. Kormanik, Ph.D. Associate Professor of Biology Department of Biology University of North Carolina at Asheville	J. Lofton N. O'Leary-Liu

Principal Investigator	Associates
Thomas H. Maren, M.D. Graduate Research Professor Department of Pharmacology University of Florida College of Medicine	A. Fine D. Rothman E. Swenson, M.D.
James Maylie, Ph.D. Assistant Professor Department of Obstetrics/Gynecology Oregon Health Sciences University	
Thomas J. McManus, M.D. Professor, Department of Cell Biology Duke University Medical Center	
David S. Miller, Ph.D. Expert Research Physiologist Laboratory of Cellular & Molecular Pharmacology NIH/NIEHS	D. Barnes, Ph.D. D. Bowen, Ph.D.
Martin Morad, Ph.D. Professor, Department of Physiology University of Pennsylvania	T. Anderson L. Cleemann, Ph.D. L. Gandia L. Sorbera
Alison Morison-Shetlar, Ph.D. Head, Molecular Biology Unit Max-Planck-Institut fuer Systemphysiology Dortmund, West Germany	D. Kinne R. Moore R. Shetlar, Ph.D.
Robert L. Preston, Ph.D. Professor, Department of Biological Sciences Illinois State University	L. Hartema S. Janssen K. McQuade S. Miller, Ph.D. K. Peterson K. Simokat
Raymond Rappaport, Ph.D. Research Scientist Mount Desert Island Biological Laboratory	B. Rappaport
J. Michael Redding, Ph.D. Assistant Professor, Department of Biology Tennessee Tech University	
J. Paul Schofield, M.D. MCR Training Fellow MRC Molecular Genetics Unit Cambridge, England	

Principal Investigator	Associates
Patricio Silva, M.D. Associate Professor of Medicine, Harvard Medical School; Associate Director, Renal Division, Beth Israel Hospital	J. Landsberg M. Taylor
Daniel J. Smith, M.D. Assistant Professor of Medicine University of Wisconsin Medical School at Madison	
Richard J. Solomon, M.D. Associate Professor of Medicine and Pharmacology New York Medical College	H. Solomon
Hilmar Stolte, M.D. Professor and Academic Director Department of Internal Medicine Hannover Medical School Hannover, West Germany	L. Emunds L. Fels, Ph.D. S. Kastner S. Piippo
David W. Towle, Ph.D. Foster G. McGaw Professor Chairperson of Biology Lake Forest College	S. Bowring M. Kordylewski
Jose A. Zadunaisky, M.D., Ph.D. Professor of Physiology and Biophysics Professor of Experimental Ophthalmology Director, Sackler Institute of Graduate Biomedical Studies New York University Medical Center	S. Cardona D. Giordano D. Roberts
William E. Zamer, Ph.D. Assistant Professor Department of Biology Lake Forest College	W. VanDorp

1 9 9 1 S E M I N A R S

Morning Transport

July 8   "Functional signatures for transport proteins: Looking backward
            toward the future" David C. Dawson, Ph.D., University of Michigan
            Medical School

July 16   "Epithelial chloride channels and the cystic fibrosis gene: The
            function junction" Raymond A. Frizzell, Ph.D., University of
            Alabama at Birmingham

July 22   "Molecular biology, transport and immunology - 'mix and match'"
            Alison Morrison-Shetlar, Ph.D., Max-Planck Institut fuer System-
            physiologie, Dortmund, Germany

July 29   "The actin cytoskeleton in the epithelial cell" Richard M. Hays,
            M.D., Albert Einstein College of Medicine, Prof. Yuri Natochin,
            M.D., Ph.D. and Rimma Parnova, Ph.D., Sechenov Institute of Evolu-
            tionary Physiology, Leningrad, USSR

August 5   "Cloning and sequencing of cardiac C type natriuretic peptide (CNP)
            and gene in the spiny dogfish shark, Squalus acanthias" John N.
            Forrest, Jr., M.D., Yale University School of Medicine, J. Paul
            Schofield, M.D. and D. Stephen C. Jones, Ph.D., MRC Molecular
            Genetics Unit, Cambridge, England

August 12   "Expression cloning of membrane transport systems" James L. Boyer,
            M.D., Yale University School of Medicine

August 19   "Progress in the Na,K,Cl-cotransporter in shark rectal gland"
            Bliss Forbush III, Ph.D., Yale University School of Medicine

August 26   "Band 3 and skate RBC volume regulation" Leon Goldstein, Ph.D.,
            Brown University

Morning Toxicology

July 19   "Mechanisms of metal toxicity in vascular smooth muscle" David H.
            Evans, Ph.D., University of Florida

July 26   "Effect of mercurials on the membrane transport of organic mole-
            cules by invertebrate red cells" Robert L. Preston, Ph.D., Illi-
            nois State University

August 9   "Sodium-cotransport systems and cadmium"   Rolf Kinne, M.D., Ph.D.,
            Max-Planck-Institut fuer Systemphysiologie, Dortmund, Germany

Noon

July 5   "Experimental analysis of animal cell division" Raymond Rappaport,
            Ph.D., Mount Desert Island Biological Laboratory

"Stimulation of chloride secretion in the shark rectal gland: Role of atrial natriuretic peptide" Karl Karnaky, Jr., Ph.D., Medical University of South Carolina

July 12   "Regulation of spermatogenesis in shark" J. Michael Redding, Ph.D., Tennessee Tech University

Several short presentations by the following: Gary W. Conrad, Ph.D., Kansas State University; Gregg A. Kormanik, Ph.D., University of North Carolina at Asheville; David S. Miller, Ph.D., NIH/NIEHS; Robert L. Preston, Ph.D., Illinois State University; Richard Solomon, M.D., Harvard Medical School and New England Deaconess Hospital

July 19   "Testing the free radical theory of aging" Barbara Kent, Ph.D., Mount Desert Island Biological Laboratory

"Toward an alternate cytochrome oxidase: The photochemical action spectrum for reversal of co-inhibition in skate gastric mucosal cells" George W. Kidder III, Illinois State University

"Regulation of spermatogenesis: The shark testis model" Gloria V. Callard, Ph.D., Boston University

July 26   "Electrogenic Na/H antiporter: Beginning Molecular studies" David W. Towle, Ph.D., Lake Forest College

"Molluscan models of mammalian reperfusion injury: Free radicals revisited", James A. Dykens, Ph.D., Grinnell College

August 2   "Evolution and regulation of a calcium binding protein gene" Nelson B. Horseman, Ph.D., University of Cincinnati Medical School

Evening

July 3   "Cell to cell communications in ovarian development" John J. Eppig, Ph.D., The Jackson Laboratory, Bar Harbor, Maine

July 10   "Directional cell movements during embryogenesis" J.P. Trinkaus, Ph.D., Yale University

July 17   "Biological motors and embryo development" Dan Kiehart, Ph.D., Harvard University

July 24   "Estrogen regulation of uterine proto-oncogene expression" George M. Stancel, Ph.D., The University of Texas Medical School at Houston

July 31   (William L. Doyle Lecture) "Adenosine receptors and signaling in the nephron" William S. Spielman, Ph.D., Michigan State University

August 5   "The 'new' endocrinology" Howard A. Bern, Ph.D., University of California at Berkeley

August 7   THE TENTH WILLIAM B. KINTER MEMORIAL LECTURE. "Environmental
            health: Global and local issues" David P. Rall, M.D., Ph.D.,
            NIH/NIEHS

August 14   "Exploring the life history of the sodium pump" Douglas M. Fam-
            brough, Ph.D., The Johns Hopkins University

Special Seminars

July 18   "Islands: Crucibles of Evolution" Prof. William H. Amos, Marine
            Biologist, Nature Photographer and Science Writer, Lyndonville,
            Vermont. Public lecture to enhance MDIBL's educational program for
            high school and undergraduate students funded by the Pew Founda-
            tion, Burroughs Wellcome Fund and the Hearst Foundation

July 18   CURRENT RUMOR IN MOLECULAR TECHNIQUES - Topic: "Making cDNA li-
            braries using items available at Don's Shop & Save" - Informal
            discussion. D. Stephen C. Jones, Ph.D., MRC Molecular Genetics
            Unit, Cambridge, England

July 23   CURRENT RUMOR IN MOLECULAR TECHNIQUES - "Pulling a rabbit
            out of a shark: Uses of nucleic acid homology and PCR" J. Paul
            Schofield, M.D., MRC Molecular Genetics Unit, Cambridge, England

July 30   CURRENT RUMOR IN MOLECULAR TECHNIQUES - Topic: "Egg on my face:
            Life with the xenopus oocyte" - Round table discussion. Leader,
            David S. Miller, Ph.D., NIH/NIEHS

August 8   "Future Science: Young Scientists at work today!" Local and
            national high school and college students participating in exciting
            hands-on training programs at the MDIBL and The Jackson Laboratory
            describe their research projects.

1991 PUBLICATIONS

Ballatori, N. Mechanisms of metal transport across liver cell plasma membranes. Drug Metabolism Reviews, 23(1&2), 83-132, 1991.

Brown, O.M. and J.S. Andrake. Cadmium inhibits stimulus-response coupling in skate (Raja erinacea) electric organ. Comp. Biochem. Physiol. - Part C - Comp. Pharm. Tox., 1992, in press.

Callard, G.V. Reproduction in male elasmobranch fishes. In: Comparative Physiology: "Oogenesis, Spermatogenesis, and Reproduction", R.K.H. Kinne, E. Kinne-Saffran and K.W. Beyenback (Eds.), Karger, Basel, Vol. 10, pp. 104-154, 1991.

Callard, G.V. Spermatogenesis. In: Vertebrate Endocrinology: Fundamentals and Medical Implications, P. Pang and M. Schreibman (Eds.), Academic Press, NY, Vol. 4, Part A., pp. 303-341, 1991.

Callard, I.P., K. Etheridge, G. Giannoukos, T. Lamb and L.E. Perez. The role of steroids in reproduction in female elasmobranchs and reptiles. J. Steroid Biochem. Molec. Biol., 1991.

Callard, I.P., L.A. Fileti, L.E. Perez, L.A. Sorbera and T.J. Koob. Ovarian and reproductive tract regulatory mechanisms in elasmobranchs. Environmental Biology of Fishes, 1992, in press.

Claiborne, J.B. and D.H. Evans. Acid-base balance and ion transfers in the spiny dogfish (Squalus acanthias) during hypercapnia: a role for ammonia excretion. J. Exp. Zool. 261:9-17, 1992, in press.

Conrad, A.H., A.Q. Paulsen and G.W. Conrad. The role of microtubules in contractile ring function. J. Exptl. Zool., 1991, in press.

Dubois, W. and G.V. Callard. Culture of intact Sertoli/germ cell units and isolated Sertoli cells from Squalus testis: (I) Evidence of stage-related functions in vitro. J. Exp. Zool. 258:359-372, 1991.

Ecay, T.W. and J.D. Valentich. Chloride secretagogues stimulate inositol phosphate formation in shark rectal gland tubules cultured in suspension. J. Cell. Physiol. 146:407-416, 1991.

Evans, D.H. Rat atriopeptin dilates vascular smooth muscle of the ventral aorta from the shark (Squalus acanthias) and the hagfish (Myxine glutinosa). J. Exp. Biol. 157:551-555, 1991.

Evans, D.H. and Y.A. Takei. A putative role for natriuretic peptides in fish osmoregulation. News Physiol. Sci., 1991, in press.

Evans, D.H. Evidence for the presence of A_1 and A_2 adenosine receptors in the ventral aorta of the dogfish shark, Squalus acanthias. J. Comp. Physiol., 1992, in press.

Fileti, L.A. and I.P. Callard. Regulation of ovarian steroidogenesis in the little skate, Raja erinacea. Serono Symposia. In: Signalling mechanisms and gene expression in the ovary, G. Gibori (Eds.), Springer-Verlag, Inc. NY, pp. 400-404, 1991.

Forbush, B. III, M. Haas and C. Lytle. Na-K-Cl cotransport in the shark rectal gland. I. Regulation in the intact perfused gland. Am. J. Physiol., 1992, in press.

Karnaky, K.J., Jr. Teleost osmoregulation: structure of tight junctions in relation to environmental salinity. In: The tight junction, M. Cereijido (Ed.), CRC Press, Boca Raton, pp. 175-185, 1991.

Karnaky, K.J., Jr., J.D. Valentich, M.G. Currie, W.F. Oehlenschlager and M.P. Kennedy. Atriopeptin stimulates chloride secretion by cultured shark rectal gland epithelial cells. Amer. J. Physiol. (Cell Physiol.), 260:C1125-C1130, 1991.

Kastner, S., M.F. Wilks, M. Soose, P.H. Bach and H. Stolte. Metabolic studies on isolated rat glomeruli. A valuable tool to investigate glomerular damage. In: Nephrotoxicity: Mechanisms, early diagnosis and therapeutic management, P.H. Bach, N.J. Gregg, M.F. Wilks, L. Delacruz (Eds.), Marcel Dekker, New York, 467-473, 1991.

Kastner, S., M.F. Wilks, W. Gwinner, M. Soose, P.H. Bach and H. Stolte. Metabolic heterogeneity of isolated cortical and juxtamedullary glomeruli in Adriamycin nephrotoxicity. Renal Physiol. Biochem., 14:48-54, 1991.

Kelley, G.G., O.S. Aassar and J.N. Forrest, Jr. Endogenous adenosine is an autacoid feedback inhibitor of chloride transport in the shark rectal gland. J. Clin. Invest., 88: (6), 1991, in press.

Kidder, G.W. III. Effects of luminal osmolarity on gastric acid secretion in the little skate, Raja erinacea. J. Comp. Physiol. 161:323-326., 1991.

Koob, T.J. Deposition and binding of calcium and magnesium in egg capsules of Raja erinacea Mitchill during formation and tanning in utero. Copeia, (2): 339-347, 1991.

Koob, T.J. and I.P. Callard. Reproduction in female elasmobranchs. In: Comparative Physiology: "Oogenesis, spermatogenesis and reproduction", R.K.H. Kinne (Ed.), Karger Publishing, Basel, Vol. 10, pp. 155-209, 1991.

Lytle, C. and B. Forbush, III. Na-K-Cl cotransport in the shark rectal gland. II. Regulation in isolated tubules. Am. J. Physiol. 1992, in press.

Maren, T.H., A. Fine, E.R. Swenson and D. Rothman. Renal acid-base physiology in the marine teleost, Myoxocephalus octodecimspinosus (long-horned sculpin). Am. J. Physiol. (Renal Fluid and Electrolyte Physiol.), 1991, in press.

Moran, W.M. and J.D. Valentich. Cl⁻ secretion by cultured shark rectal gland cells. II. Effects of forskolin on cellular electrophysiology. Am. J. Physiol. 260 (Cell Physiol. 29):C824-C831, 1991.

Perez, L.E., M. Fenton and I.P. Callard. Vitellogenins - homologs of mammalian apolipoproteins? Comp. Physiol. Biochem., 1991, in press.

Pritchard, J.B. and D.S. Miller. Comparative insights into the mechanisms of renal organic anion and cation secretion. Amer. J. Physiol. 261:, 1991, in press.

Rappaport, R. Cytokinesis. In: "Oogenesis, Spermatogenesis and Reproduction", R.K.H. Kinne (Ed.), Karger, Basel, pp. 1-36., 1991.

Rappaport, R. Enhancement of aster-induced furrowing activity by a factor associated with the nucleus. J. Exp. Zool. 257:87-95, 1991.

Reese, J.C. and I.P. Callard. Characterization of a specific estradiol receptor in the oviduct of the little skate, Raja erinacea. Gen. Comp. Endocrinol. 84:170-181, 1991.

Schofield, J.P., D.S.C. Jones and J.N. Forrest, Jr. Identification of C-type natriuretic peptide in heart of spiny dogfish shark (Squalus acanthias). Amer. J. Physiol., 261:F734-F739, 1991.

Sellinger, M., N. Ballatori and J.L. Boyer. Mechanism of mercurial inhibition of sodium-coupled alanine uptake in liver plasma membrane vesicles from Raja erinacea. Toxicol. Appl. Pharmacol., 107:369-376, 1991.

Silva, P., R. Solomon and F.H. Epstein. Perfusion of shark rectal gland. In: Methods in Enzymology, Academic Press, 192:754-766, 1991.

Smith, D.J. and S.A. Ploch. Isolation of Raja erinacea basolateral liver plasma membranes: characterization of lipid composition and fluidity. J. Exp. Zoology, 258:189-195, 1991.

Soder, O., V. Syed, B. Froysa, G.V. Callard, J. Toppari, P. Pollanen, M. Parvinen and E.M. Ritzen. Production and secretion of an interleukin-1-like factor correlates with spermatogonial DNA synthesis in the rat seminiferous tubule. Int. J. Androl., 14:223-231, 1991.

Tsang, P. and I.P. Callard. Regulation of in vitro ovarian steroidogenesis in the viviparous shark, Squalus acanthias. J. Exp. Zool., 1991, in press.

Ubels, J.L. and T.B. Osgood. Inhibition of corneal epithelial cell migration by cadmium and mercury. Bull. Environ. Contam. Toxicol. 46:230-236, 1991.

Valentich, J.D. Primary cultures of shark rectal gland epithelial cells: A model for hormone-sensitive chloride transport. J. Tiss. Cult. Meth., 1991, in press.

Valentich, J.D. and J.N. Forrest, Jr. Cl-secretion by cultures shark rectal gland cells. I. Transepithelial transport. Am. J. Physiol., 260 (Cell Physiol. 29):C813-C823, 1991.

Wright, B.D., J.H. Henson, K.P. Wedaman, P.J. Willy, J.N. Morand and J.M. Scholey. Subcellular localization and sequence of sea urchin kinesin heavy chain: Evidence for its association with membranes in the mitotic apparatus and interphase cytoplasm. J. Cell Biol., 113:817-833, 1991.

Ziyadeh, F.N., J.W. Mills and A. Kleinzeller. Hypotonicity and cell volume regulation in shark rectal gland: role of organic osmolytes and F-actin. Am. J. Physiol., 1991, in press.

SPECIES INDEX